MATHEMATICS RESEARCH DEVELOPMENTS

UNDERGRADUATE RESEARCH AT CAMERON UNIVERSITY ON ITERATIVE PROCEDURES IN BANACH AND OTHER SPACES

MATHEMATICS RESEARCH DEVELOPMENTS

Additional books and e-books in this series can be found on Nova's website under the Series tab.

MATHEMATICS RESEARCH DEVELOPMENTS

UNDERGRADUATE RESEARCH AT CAMERON UNIVERSITY ON ITERATIVE PROCEDURES IN BANACH AND OTHER SPACES

IOANNIS K. ARGYROS
AND
SAMUNDRA REGMI

nova
science publishers
New York

Additional color graphics may be available in the e-book version of this book.

Library of Congress Cataloging-in-Publication Data

ISBN: 978-1-53616-058-1
Library of Congress Control Number: 2019947073

Published by Nova Science Publishers, Inc. † New York

The first author dedicates this book to his beloved parents,
***Anastasia**, and **Konstantinos**,*
*wife **Diana**, children **Michael, Gus, Christopher**, and **Stacey**.*

The second author dedicates this book to his parents,
Madhu Kumari**, and **Moti Ram,
*siblings **Shanta, Srijana, Sarita, Santosh,** and **Sarmila,***
*nephews **Bishwas,** and **Saimon,** nieces **Binisha,** and **Asmini**.*

Contents

Preface **xiii**

1 Majorizing Sequences for Iterative Methods with Applications **1**
1. Introduction 1
2. Semi-Local Convergence 2
3. Numerical Example 10
References 13

2 On a Fast Three Step Method for Solving Equations under Weak Conditions **15**
1. Introduction 15
2. Local Convergence Analysis 16
3. Numerical Examples 25
References 29

3 Ball Convergence of Two Derivative Free Iterative Methods for Solving Equations under Weak Conditions **31**
1. Introduction 31
2. Local Convergence 33
3. Numerical Examples 47
References 49

4 Quasi-Newtonian Iterative Processes **51**
1. Introduction 51
2. Local Convergence 52

3. Semi-Local Convergence 56
4. Numerical Example 61
References 63

5 Extended Ball Convergence for a Chebyshev-Halley Family of Methods with One Parameter 65
1. Introduction 65
2. Local Convergence 66
3. Numerical Example 78
References 81

6 Extending the Applicability of Newton's Method under the Second Fréchet-Derivative: Case I 83
1. Introduction 83
2. Semi-Local Convergence Analysis 83
3. Local Convergence Analysis 89
References 95

7 Extending the Applicability of Newton's Method under the k-th (k $\geq$ 2) Fréchet Derivaitive: Case II 97
1. Introduction 97
2. Semi-Local Convergence Analysis 97
References 103

8 Convergence under w-Conditions and the k-Fréchet Derivatives: Case III 105
1. Introduction 105
2. Semi-Local Convergence Analysis 105
3. Local Convergence Analysis 111
References 119

9 Convergence under w-Conditions on the Second Derivative: Case IV 121
1. Introduction 121
2. Semi-Local Convergence Analysis 121
3. Local Convergence Analysis 127
References 131

10 Convergence under Lipschitz Conditions: Case V 133
1. Introduction 133
2. Semi-Local Convergence Analysis 133
3. Local Convergence Analysis 139
References 143

11 Convergence under w-Lipschitz Conditions: Case VI 145
1. Introduction 145
2. Semi-Local Convergence Analysis 145
3. Local Convergence Analysis 151
References 159

12 Convergence of an Eight Order Four Step Method 161
1. Introduction 161
2. Local Convergence 162
References 169

13 A Fast Three Step Method 171
1. Introduction 171
2. Local Convergence 172
References 181

14 Convergence of a Method Using Derivatives and Divided Differences 183
1. Introduction 183
2. Local Convergence 184
References 191

15 Convergence of a 2k Method 193
1. Introduction 193
2. Local Convergence 194
References 203

16 Convergence of an Eighth Order Method **205**
1. Introduction 205
2. Local Convergence 206
References 213

17 Convergence for a Method Containing a Sum of Linear Operators **215**
1. Introduction 215
2. Local Convergence 216
References 223

18 Local Convergence of a Two-Step Chebyshev-Type Method **225**
1. Introduction 225
2. Local Convergence 226
References 233

19 Local Convergence of a Three Parameter Sixth Order Method **235**
1. Introduction 235
2. Local Convergence 236
References 243

20 Local Convergence for a Steffensen-Type Method of Order Eight **245**
1. Introduction 245
2. Local Convergence 246
References 253

21 Convergence of a Sixth Order Method **255**
1. Introduction 255
2. Local Convergence 256
References 265

22 Roots of Polynomials and Their Applications **267**
1. Introduction 267
2. Locating Roots of Polynomials 268
3. Applications and Examples 277
References 283

23 Hybrid High Convergence Order Iterative Methods 285
1. Introduction 285
2. Local Convergence 286
3. Numerical Examples 290
References 291

24 Hybrid Newton-Like Methods with High Order of Convergence 293
1. Introduction 293
2. Local Convergence 293
3. Numerical Examples 298
References 299

Glossary of Symbols 301

About the Authors 303

Index 305

Related Nova Publications 309

Preface

This book is an outgrowth of Undergraduate Research work conducted at Cameron University, Department of Mathematical Sciences during the period 2017-2019. The authors are Dr. Ioannis K. Argyros (Advisor) and Mr. Samundra Regmi (Mathematics, Computer Science, and Information Technology student).

Next, we present a synopsis intended to answer to some questions. The society becomes more and more complicated due to the explosion of technology and other factors. People find it increasingly difficult to adjust or understand what is going on around them. The only way to solve this problem is through decent education. This is our motivation behind this book, and that is why we wrote it. In particular, with our undergraduate research we attempt to contribute toward the solution of the aforementioned problems.

What are we doing? A plethora of problems from mathematics, physics, economics, computer science, decision making, complex iterative methods, systems, modeling and control, engineering, social and life sciences, and other disciplines can be brought in the form of a nonlinear equation defined on abstract spaces using mathematical modeling. Solutions are needed for these equation preferably in closed form. But, a closed form solution which may be a number or vector or function with physical significance can be obtained only in special cases. That is why iterative methods are used to solve such equations. These methods lead to the development of proper computer programs to obtain accurate approximate solutions.

This book presents iterative methods generating a sequence approximating solutions of equations involving Banach or Hilbert or Euclidean or other space valued operators. The book contains numerous new results on local and semi-local convergence of iterative methods converging under weaker hypotheses than in earlier works, and under the same computational cost. Hence, the applicability of iterative methods is extended.

The book is a valuable resource for researchers, practitioners, graduate students and senior undergraduate students. It can be used as a text book for a computational mathematics class and for seminars in the aforementioned disciplines.

What are we going to do after publication? We will contact not only the people listed in the dedication pages but many others to stimulate them, so they can imitate us.

What should you expect from us? We will travel to deliver talks in conferences about our book, to stress the importance of undergraduate research, and using our experience provide any needed assistance.

Ioannis K. Argyros
Samundra Regmi

Chapter 1

Majorizing Sequences for Iterative Methods with Applications

1. Introduction

In this chapter we are concerned with the problem of approximating a locally unique solution of x^* of the equation

$$F(x) = 0, \tag{1.1}$$

where, $F : D \subset X \to Y$ is a Fréchet-differentiable operator, X, Y are Banach spaces, $D \subset X$ is an open and convex set. There is a plethora of problems from different disciplines such as Applied Mathematics, Optimization, Mathematical Biology, Chemistry, Economics, Physics and also engineering that can be brought in form like equation (1.1) using mathematical modeling [1-3]. The solution x^* is needed in closed form but this can be done only in special cases. That is why researchers and practitioners rely on iterative algorithms to generate a sequence approximating x^*.

The most popular method is undoubtedly Newton's method defined for each n = 0, 1, 2, ... by

$$x_{n+1} = x_n - F'(x_n)^{-1}F(x_n), \tag{1.2}$$

where $x_0 \in$ D is an initial point.

The study of convergence of iterative algorithms is usually centered into two categories: semi-local and local convergence analysis. The semi-local convergence is based on the information around an initial point, to obtain conditions ensuring the convergence of these algorithms, while the local convergence is based on the information around a solution to find estimates of the computed radii of the convergence balls. Local results are important since they provide the degree of difficulty in choosing initial points.

In this chapter, we show how to expand the applicability of Newton's method in cases not covered before. In particular we obtain the following advantages (A) over earlier works:

(A_1) Larger convergence domain.

(A_2) Tighter error bounds on the distances $\|x_{n+1}-x_n\|, \|x_n-x^*\|$.

(A_3) More precise information on the location of the solution x^*.

The rest of the chapter contains the semi-local convergence of Newton's method in Section 2 and some examples in the concluding Section 3.

2. Semi-Local Convergence

We shall present the semi-local convergence analysis of Newton's method (1.2) in this section. First, we need some auxiliary results and definitions.

Definition 2.1 Let $\{x_n\}$ be a sequence in X. Then, a sequence $\{u_n\} \subset$ **[0,**∞) for which

$$\|x_{n+1}-x_n\| \leq u_{n+1}-u_n \text{ for each } n=0,1,\ldots \tag{2.1}$$

holds is a majorizing sequence for $\{x_n\}$.

Proposition 2.2 Let $\{u_n\}$ be a majorizing sequence for $\{x_n\}$, and suppose that

$$\lim_{n\to+\infty} u_n = u^* < \infty \text{ exists}$$

Then

$$x^* = \lim_{n \to +\infty} x_n \text{ exists}$$

and

$$\|x_n - x^*\| \leq u^* - u_n \text{ for each } n = 0, 1, 2, \ldots \tag{2.2}$$

Proof. We have in turn by (2.1) that

$$\|x_{n+m} - x_n\| \leq \sum_{j=n}^{n+m-1} \|x_{j+1} - x_j\|$$

$$\leq \sum_{j=n}^{n+m-1} (u_{j+1} - u_j) = u_{n+m} - u_n. \tag{2.3}$$

It follows from (2.3) that sequence $\{x_n\}$ is Cauchy (or complete) in a Banach space X and as such it converges to its limit point x^*. Then, estimate (2.2) follows from (2.3) by letting $m \to \infty$.

□

Definition 2.3 Let $F : D \to Y$ be a Fréchet-differentiable operator. Suppose that for some $x_0 \in D$, $F'(x_0)^{-1}$ is invertible.

(a) We say that operator F is center-Lipschitz on D, if there exists $L_0 > 0$ such that

$$\|F'(x_0)^{-1}(F'(x) - F'(x_0))\| \leq L_0 \|x - x_0\| \text{ for each } x \in D. \tag{2.4}$$

(b) We say that operator F' is Lipschitz on D, if there exists $L > 0$ such that

$$\|F'(x_0)^{-1}(F'(x) - F'(y))\| \leq L\|x - y\| \text{ for each } x, y \in D. \tag{2.5}$$

Remark 2.4 It follows from (2.4) and (2.5) that

$$L_0 \leq L \tag{2.6}$$

holds and $\frac{L_0}{L}$ can be arbitrarily small [1,2].
To continue we show that inequality (2.6) can be strict.

Example 2.5 Let $X = Y = \mathbb{R}$, $x_0 = 1$ and $D = U(x_0, 1-p)$, for some $p \in [0, \frac{1}{2})$. Define function F on D by,

$$F(x) = x^3 - p. \tag{2.7}$$

Then, using (2.4), (2,5) and (2.7), we have that

$$L_0 = 3 - p \text{ and } L = 2(2-p). \tag{2.8}$$

Notice that

$$L_0 < L \text{ for each } p \in \left[0, \frac{1}{2}\right). \tag{2.9}$$

Let us prove (2.8). First notice that, we choose $\|.\| = |\,.\,|$. Using(2.4), (2.7) and the triangle inequality we get in turn that

$$\begin{aligned}
\|F'(x_0)^{-1}(F'(x) - F'(x_0))\| &= \frac{1}{3}|3x^2 - 3x_0{}^2| \\
&= |x^2 - x_0{}^2| = |x^2 - x_0{}^2| = |x + x_0|\,|x - x_0| \\
&\leq |(x - x_0) + 2x_0||x - x_0| \\
&\leq (|x - x_0| + 2|x_0|)|x - x_0| \\
&\leq (1 - p + 2)|x - x_0| = (3 - p)|x - x_0|, \quad (2.10)
\end{aligned}$$

so we can choose $L_0 = 3 - p$, which shows the first estimate in (2.8). Similarly, but using (2.5) and (2.7), we obtain in turn that

$$\begin{aligned}
\|F'(x_0)^{-1}(F'(x) - F'(y))\| &= \frac{1}{3}|3x^2 - 3y^2| \\
&= |x^2 - y^2| = |x + y|\,|x - y| \\
&\leq \left[|x - x_0| + |y - y_0| + 2|x_0|\right]|x - y| \\
&\leq \left[1 - p + 1 - p + 2 \cdot 1\right]|x - y| \\
&= 2(2 - p)|x - y|, \quad (2.11)
\end{aligned}$$

so we can choose $L = 2(2-p)$, which completes the proof for (2.8).

Next, we present the celebrated Newton-Kantorovich theorem for solving nonlinear equations using Newton's method.

Theorem 2.6 [3] Suppose that operator F' is Lipschitz continuous on D and

$$h_{\mathbf{K}} = 2L\eta \leq 1, \tag{2.12}$$

where

$$\|F'(x_0)^{-1}F(x_0)\| \leq \eta. \tag{2.13}$$

Then, sequence $\{x_n\}$ generated by Newton's method is well defined in $U(x_0, r)$, remains in $U(x_0, r)$ for each n = 0, 1, 2, ... and converges to a unique solution x^* of equation $F(x) = 0$ in $\overline{U}(x_0, r)$, where

$$r = \frac{1 - \sqrt{1 - 2L\eta}}{L}. \tag{2.14}$$

Moreover, the following error bounds hold

$$\|x_{n+1} - x_n\| \leq t_{n+1} - t_n \tag{2.15}$$

and

$$\|x_n - x^*\| \leq r - t_n, \tag{2.16}$$

where

$$t_0 = 0, t_1 = \eta$$

$$t_{n+2} = t_{n+1} + \frac{L(t_{n+1} - t_n)^2}{2(1 - Lt_{n+1})} \text{ for each } n = 0, 1, 2, \ldots \tag{2.17}$$

Remark 2.7 Estimate (2.12) is the famous for its simplicity and clarity sufficient convergence criterion for Newton's method due to Kantorovich (Nobel prize in (Mathematical) Economics). However, this criterion may not be satisfied. Returning back to the academic and motivation Example 2.5, we see that $\eta = \frac{1}{3}(1-p)$.
Therefore, (2.12) becomes

$$h_{\mathbf{K}} = \frac{4}{3}(1-p)(2-p) > 1 \text{ for each } p \in \left[0, \frac{1}{2}\right). \tag{2.18}$$

Hence, there is no guarantee that Newton's method converge to $x^* = \sqrt[3]{p}$.
Next, we show how to replace the Kantorovich convergence criterion (2.12) by

a weaker one, so that Newton's method will converge to x^* at least for some $p \in \left[0, \frac{1}{2}\right)$.
To achieve this, first we need an auxiliary result for majorizing sequences.

Lemma 2.8 Let $L_0 > 0$, $L > 0$ and $\eta \geq 0$.
Suppose that

$$h_{\mathbf{A}} = 2K\eta \leq 1, \tag{2.19}$$

where

$$K = \frac{1}{4}(L + L_0 + \sqrt{L_0^2 + 8L_0L}).$$

Then, scalar sequence $\{s_n\}$ defined by $s_0 = 0$, $s_1 = \eta$,

$$s_{n+2} = s_{n+1} + \frac{L(s_{n+1} - s_n)^2}{2(1 - L_0 s_{n+1})} \text{ for each } n = 0, 1, 2, \ldots \tag{2.20}$$

is well defined, nondecreasing bounded from above by

$$s^{**} = \frac{\eta}{1 - \alpha} \tag{2.21}$$

and convergent to its unique least upper bound s^* which satisfies

$$\eta \leq s^* \leq s^{**}, \tag{2.22}$$

where

$$\alpha = \frac{2L}{L + \sqrt{L^2 + 8L_0L}}. \tag{2.23}$$

Proof. We shall use mathematical induction to show

$$0 \leq \frac{L(s_{k+1} - s_k)}{2(1 - L_0 s_{k+1})} \leq \alpha. \tag{2.24}$$

Using, (2.19) and (2.20) it is simple algebra to show that estimate (2.24) holds for $k = 0$. Then, in view of (2.21) and (2.24) for $k = 0$, we have that

$$0 \leq s_2 - s_1 \leq \alpha(s_1 - s_0) = \alpha\eta,$$

so,

$$0 \leq s_2 - s_1 \leq \alpha\eta$$

and

$$s_2 \leq s_1 + \alpha\eta = (1+\alpha)\eta = \frac{1-\alpha^2}{1-\alpha}\eta < \frac{\eta}{1-\alpha} = s^{**}.$$

Suppose that

$$0 \leq s_{k+1} - s_k \leq \alpha^k \eta \tag{2.25}$$

and

$$s_{k+1} \leq \frac{1-\alpha^{k+1}}{1-\alpha}\eta < s^{**}. \tag{2.26}$$

It suffices to show

$$0 \leq \frac{L\alpha^k \eta}{2\left[1 - L_0 \frac{1-\alpha^{k+1}}{1-\alpha}\eta\right]} \leq \alpha,$$

or

$$L\alpha^k \eta \leq 2\alpha - 2L_0 \frac{1-\alpha^{k+1}}{1-\alpha}\eta,$$

or

$$L\alpha^k \eta + 2\alpha L_0 \frac{1-\alpha^{k+1}}{1-\alpha}\eta - 2\alpha \leq 0,$$

or

$$L\alpha^{k-1}\eta + 2L_0(1+\alpha+\cdots+\alpha^k)\eta - 2 \leq 0 \tag{2.27}$$

Estimate (2.27) motivates us to introduce recurrent functions δ_k defined on interval [0,1)

$$\delta_k(t) = L\eta t^{k-1} + 2L_0(1+t+\cdots+t^k)\eta - 2. \tag{2.28}$$

We need a relationship between two consecutive functions δ_k.
We can write by (2.28) that

$$\begin{aligned}
\delta_{k+1}(t) &= L\eta t^k + 2L_0(1+t+\cdots+t^{k+1})\eta - 2 - \delta_k(t) + \delta_k(t) \\
&= L\eta t^k + 2L_0(1+t+\cdots+t^{k+1})\eta - 2 - L\eta t^{k-1} \\
&\quad - 2L_0(1+t+\cdots+t^{k+1})\eta + 2 + \delta_k(t) \\
&= \delta_k(t) + g(t)t^{k-1}\eta,
\end{aligned}$$

where $g(t) = 2L_0 t^2 + Lt - L$.
Notice that $g(\alpha) = 0$ and $\alpha \in (0,1)$, where α is given in (2.23). That is we showed

$$\delta_{k+1}(t) = \delta_k(t) + g(t)t^{k-1}\eta. \tag{2.29}$$

Estimate (2.29) can be rewritten as

$$\delta_k(\alpha) \leq 0. \tag{2.30}$$

Notice that

$$\delta_{k+1}(\alpha) = \delta_k(\alpha), \tag{2.31}$$

since $g(\alpha) = 0$.
Define function $\delta_\infty(t)$ on [0,1] by

$$\delta_\infty(t) = \lim_{k \to +\infty} \delta_k(t). \tag{2.32}$$

Then, in view of (2.30)-(2.32), we can show instead of (2.30) that

$$\delta_\infty(\alpha) \leq 0. \tag{2.33}$$

Using (2.33) and (2.34), we get

$$\delta_\infty(\alpha) = 2\Big[\frac{L_0\eta}{1-\alpha} - 1\Big]. \tag{2.34}$$

By (2.33) and (2.34), we must show that

$$\frac{L_0\eta}{1-\alpha} \leq 1, \tag{2.35}$$

which is true by (2.19). The induction for (2.25) and (2.26) is completed. Then, notice by (2.25) and (2.26) that sequence $\{s_k\}$ is nondecreasing and bounded above by s^{**} and as such it converges to its unique least upper bound s^* which satisfies s^*. □

Then, we have our weaker Newton-Kantorovich Theorem.

Theorem 2.9 Suppose that operator F' satisfies (2.4), (2.5) and (2.12). Then, sequence $\{x_n\}$ generated by Newton's method is well defined in $U(x_0, s^*)$, remains in $U(x_0, s^*)$ for each n = 0, 1, 2, ... and converges to a unique solution x^* of equation $F(x) = 0$ in $\bar{U}(x_0, s^*)$. Moreover, the following error bounds hold

$$\|x_{n+1} - x_n\| \leq s_{n+1} - s_n$$

and

$$\|x_n - x^*\| \leq s^* - s_n.$$

Remark 2.10 Notice that from (2.12),(2.19) and

$$K \leq L \tag{2.36}$$

and

$$h_{\mathbf{K}} \leq 1 \Longrightarrow h_{\mathbf{A}} \leq 1, \tag{2.37}$$

but not necessarily vice versa, unless if $L_0 = L$. That is we have extended the convergence domain for Newton's method. Moreover, a simple induction argument shows that

$$s_n \leq t_n, \tag{2.38}$$

$$0 \leq s_{n+1} - s_n \leq t_{n+1} - t_n \tag{2.39}$$

and

$$s^* = \lim_{n \to +\infty} s_n \leq r = \lim_{n \to +\infty} t_n. \tag{2.40}$$

We also have

$$\frac{h_{\mathbf{A}}}{h_{\mathbf{K}}} \to \frac{1}{4} \text{ as } \frac{L_0}{L} \to 0. \tag{2.41}$$

Implication (2.41) shows that we have extended the convergence domain at most four times. Items (2.37) - (2.40) justify the advantages stated at the introduction of this chapter. It is worth noticing that these advantages are obtained under the same computational cost, since in practice the computation of L requires the computation of L_0 and K as special cases.

Remark 2.11 Returning back to Example 2.5, it is simple algebra to show that

$$h_{\mathbf{A}} = 2K\eta \leq 1 \text{ for each } p \in [0.450339002, 0.5]. \tag{2.42}$$

Hence, Newton's method converges to x^* under our assumptions.
Numerous other examples where $L_0 < L$ $(i.e.\ K < L)$ can be found in [1,2].

3. Numerical Example

Example 3.1 Let $X = Y = C[0,1]$, the space of continuous functions defined in [0,1] equipped with the max norm. Let $D^* = \{x \in C[0,1] \mid \|x\| \leq R\}$, such that $R > 0$ and F is defined on D^* and given by

$$F(x)(s) = x(s) - f(s) - \xi \int_0^1 Q(s,t)x(t)^3 dt, \quad x \in C[0,1], \; s \in [0,1],$$

where $f \in C[0,1]$ is a given function, ξ is a real constant and the kernel Q is the Green's function. In this case, for each $x \in D^*$, $F'(x)$ is a linear operator defined on D^* by the following expression.

$$[F(x)(v)](s) = v(s) - 3\xi \int_0^1 G(s,t)x(t)^2 v(t) dt, \quad v \in C[0,1], \; s \in [0,1].$$

If we choose $x_0(s) = f(s) = 1$, it follows $\|I - F'(x_0)\| \leq \frac{3|\xi|}{8}$. Thus, if $|\xi| \leq \frac{8}{3}$, $F'(x_0)^{-1}$ is defined and

$$\|F'(x_0)^{-1}\| \leq \frac{8}{8-3|\xi|}, \quad \|F(x_0)\| \leq \frac{|\xi|}{8}, \quad \eta = \|F'(x_0)^{-1}F(x_0)\| \leq \frac{|\xi|}{8-3|\xi|}.$$

Choosing $\xi = 1.00$ and $\lambda = 3$, we have

$$\eta = 0.2, L = 3.8, L_0 = 2.6 \text{ and } K = 2.88.$$

Using these values we obtain that condition (2.12) is not satisfied, since

$$h_{\mathbf{K}} = 1.52 > 1,$$

but condition (2.19) is satisfied, since

$$h_{\mathbf{A}} = 0.985779\cdots < 1.$$

Then, the convergence of Newton's method follows by Theorem 2.9.

Moreover, there is no guarantee that Newton's method converges under the Kantorovich condition. Hence, we extended the applicability of Newton's method.

Example 3.2 Set $D = U(x_0, 1-p)$ for $p \in [0,1)$. Notice that for $p = 0.5000000000000001$ both (2.12) and (2.19) hold. So, we can compare corresponding majorizing sequences given by (2.17) and (2.20) in the following table.

Comparison Table 1.

n	s_n	t_n	$s_{n+1} - s_n$	$t_{n+1} - t_n$
0	0	0	0.166666666666667	0.166666666666667
1	0.16666666666666	0.166666666666667	0.0714285714285714	0.0833333333333333
2	0.238095238095238	0.25	0.0189075630252101	0.0416666666666666
3	0.257002801120448	0.291666666666667	0.00150001234578062	0.0208333333333332
4	0.258502813466229	0.3125	9.54098279665194E-06	0.0104166666666664
5	0.258512354449025	0.322916666666666	3.86028153886997E-10	0.00520833333333282
6	0.258512354835054	0.328124999999999	0	0.00260416666666563
7	0.258512354835054	0.330729166666665	0	0.00130208333333126
8	0.258512354835054	0.332031249999996	0	0.000651041666662522
9	0.258512354835054	0.332682291666658	0	0.000325520833325044
10	0.258512354835054	0.333007812499983	0	0.000162760416650087
11	0.258512354835054	0.333170572916633	0	8.13802083001747E-05
12	0.258512354835054	0.333251953124934	0	4.06901041003493E-05
13	0.258512354835054	0.333292643229034	0	2.03450519506987E-05
14	0.258512354835054	0.333312988280985	0	1.01725257763974E-05
15	0.258512354835054	0.333323160806761	0	5.08626249029476E-06
16	0.258512354835054	0.333328247069251	0	2.54313044933951E-06
17	0.258512354835054	0.333330790199701	0	1.27156363305403E-06
18	0.258512354835054	0.333332061763334	0	6.35778633295558E-07
19	0.258512354835054	0.333332697541967	0	3.17882950295889E-07
20	0.258512354835054	0.333333015424917	0	1.58928743332343E-07
21	0.258512354835054	0.333333174353661	0	7.94389151681507E-08
22	0.258512354835054	0.333333253792576	0	3.96686016257419E-08
23	0.258512354835054	0.333333293461178	0	1.97330433104881E-08
24	0.258512354835054	0.333333313194221	0	9.66758090248376E-09
25	0.258512354835054	0.333333322861802	0	4.46267672815637E-09
26	0.258512354835054	0.333333327324478	0	1.65717795042042E-09
27	0.258512354835054	0.333333328981656	0	3.15537984629799E-10
28	0.258512354835054	0.333333329297194	0	1.23340782032244E-11
29	0.258512354835054	0.333333329309528	0	1.89293025698589E-14

The computational cost provided by our method is always less or equal, at the worst case, for Newton's method than the one given in [3]. Thus, observing the table, we can conclude that our method converges with less iteration providing the larger convergence domain following (2.38) and (2.39).

References

[1] Argyros, I.K., *Convergence and Applications of Newton-type Iterations*, Springer Verlag, Berlin, Germany, 2008.

[2] Argyros, I.K. and Magreñán, A.A., *A Contemporary Study of Iterative Methods*, Academic Press, Elsevier, San Diego, CA, USA, 2017.

[3] Kantorovich, L.V. and Akilov, G.P., *Functional Analysis in Normed Spaces*, Pergamon Press, New York, 1964.

Chapter 2

On a Fast Three Step Method for Solving Equations under Weak Conditions

1. Introduction

Let $F : D \subset S \to S$, be a continuously differentiable function, where $S = \mathbb{R}$ of $S = \mathbb{C}$. There is a plethora of problems from all disciplines that can be brought in the form of nonlinear equation like

$$F(x) = 0, \tag{1.1}$$

using mathematical modeling [1-13]. It is then important to find a locally unique solution x^* of equation (1.1). The solution is needed in closed form but this can be achieved only in special cases. That explains why researchers and practitioners resort to the production of iterative methods to generate a sequence converging to x^* under certain conditions. Many methods have been introduced of higher order of convergence[1-13]. In particular, we revisit the three step method defined for each $n = 0, 1, 2, \dots$ by

$$y_n = x_n - \frac{F(x_n)}{F'(x_n)},$$

$$z_n = x_n - \left(1 + \frac{F(y_n)\Big(F(x_n)+(\alpha-1)F(y_n)\Big)}{F(x_n)\big(F(x_n)-F(y_n)\big)}\right)\frac{F(x_n)}{F'(x_n)} \tag{1.2}$$

$$x_{n+1} = z_n - \frac{F(z_n)}{F'(x_n)+F''(x_n)(z_n-x_n)},$$

where $x_0 \in D$ is an initial point and $\alpha \in \mathbb{R}$ is a parameter. Method (1.2) was introduced in [8] where the eighth order of convergence was shown using hypotheses up to eight derivative although these derivatives do not appear in the method. This limits the applicability of the method.

As an academic example, consider function f defined on the interval $D = [\frac{-1}{2}, \frac{3}{2}]$ by

$$f(x) = \begin{cases} x^3 \ln x^2 + x^5 - x^4, & x \neq 0 \\ 0, & x = 0. \end{cases}$$

Then, we see that $f'''(x)$ is unbounded on D. Hence, the results in [8] cannot be used.

In the present chapter, we show the convergence of the method using hypotheses until the second derivative. Moreover, we provide computable error bounds $\|x_n - x^*\|$ as well as a uniqueness of the solution result.

The rest of the chapter contains the local convergence analysis of method (1.2) in Section 2, whereas the numerical examples appear in Section 3.

2. Local Convergence Analysis

It is convenient for the local convergence analysis that follows to introduce some functions and parameters. Let $\alpha \in S$ and let $W_0 : [0,\infty) \to [0,\infty)$ be a continuous and increasing function with $W_0(0) = 0$. Suppose that equation

$$W_0(t) = 1. \tag{2.1}$$

has at least one positive solution. Denote by r_0 the smallest such solution. Let $W, V_0, V : [0, r_0) \to [0,\infty)$ be continuous and increasing functions. Define functions g_1, h_1, p and h_p on the interval $[0, r_0)$ by

$$g_1(t) = \frac{\int_0^1 W\Big((1-\theta)t\Big)d\theta}{1 - W_0(t)}$$

$$h_1(t) = g_1(t) - 1,$$

$$p(t) = \int_0^1 W_0(\theta t)d\theta + g_1(t)\int_0^1 V_0\Big(\theta g_1(t)t\Big)d\theta$$

and

$$h_p(t) = p(t) - 1.$$

We have that $h_1(0) = h_p(0) = -1$ and $h_1(t) \to \infty, h_p(t) \to \infty$ as $t \to r_0^-$. It then follows from the intermediate value theorem that functions h_1 and h_p have zeros in the interval $(0, r_0)$.
Denote by r_1, r_p, respectively the smallest such zeros. Moreover, define functions g_2, h_2, q, h_q on the interval $[0, \bar{r}_p)$ by

$$g_2(t) = g_1(t) + \frac{\Big[\int_0^1 V_0(\theta t)d\theta + |\alpha - 1|\mu\Big]\mu t}{\Big(1 - W_0(t)\Big)(1 - p(t))},$$

$$h_2(t) = g_2(t) - 1,$$

$$q(t) = W_0(t) + V(t)\Big(1 + g_2(t)\Big)t$$

and

$$h_q(t) = q(t) - 1,$$

where,

$$\mu = g_1(t)\int_0^1 V_0\Big(\theta g_1(t)t\Big)d\theta, \text{ and } \bar{r}_p = \min\{r_n, r_p\}.$$

We get that $h_2(0) = h_q(0) = -1$ and $h_2(t) \to \infty, h_q(t) \to \infty$ as $t \to \bar{r}_p^-$.
Denote by r_2, r_q, respectively the smallest zeros of functions h_2 and h_q on the

interval $(0, r_p)$. Furthermore, define functions g_3 and h_3 on the interval $[0, r_q)$ by

$$g_3(t) = g_2(t) + \frac{g_2(t) \int_0^1 V_0\left(\theta g_2(t) t\right) d\theta}{1 - q(t)}$$

and

$$h_3(t) = g_3(t) - 1,$$

where

$$\bar{r}_q = \min\{\bar{r}_p, r_q\}.$$

We obtain that $h_3(0) = -1$ and $h_3(t) \to \infty$ as $t \to \bar{r}_q^-$. Denote by r_3 the smallest zero of function h_3 in the interval $(0, r_q)$. Define the radius of convergence r by

$$r = \min\{r_i\}, \text{ for } i = 1,2,3. \tag{2.2}$$

Then, we have that for each $t \in [0, r)$

$$0 \leq g_i(t) < 1, \tag{2.3}$$

$$0 \leq p(t) < 1, \tag{2.4}$$

and

$$0 \leq q(t) < 1. \tag{2.5}$$

Let $U(\lambda, \rho), \overline{U}(\lambda, \rho)$ stand, respectively for the open and closed balls in S with center $\lambda \in S$ and of radius $\rho > 0$.

Next, we present the local convergence analysis of method (1.2) using the preceding notation.

Theorem 2.1 Let $F : D \subset S \to S$ be a twice-differentiable functions. Suppose that there exist $x^* \in D$, function $W_0 : [0, \infty) \to [0, \infty)$ continuous and nondecreasing with $W_0(0) = 0$ such that for each $x \in D$.

$$F(x^*) = 0, F'(x^*) \neq 0, \tag{2.6}$$

and

$$\|F'(x^*)^{-1}\left(F'(x) - F'(x^*)\right)\| \leq W_0(\|x - x^*\|). \tag{2.7}$$

Set $D_0 = D \cap U(x^*, r_0)$ where r_0 is defined in (2.1). Moreover, suppose that there exist $\alpha \in S$, functions $W, V_0, V : [0, r_0) \to [0, \infty)$ continuous and nondecreasing with $W(0) = 0$, such that for each $x, y \in D_0$.

$$\|F'(x^*)^{-1}\Big(F'(x) - F'(y)\Big)\| \le W(\|x - y\|), \tag{2.8}$$

$$\|F'(x^*)^{-1}F'(x)\| \le V_0(\|x - x^*\|), \tag{2.9}$$

$$\|F'(x^*)^{-1}F''(x)\| \le V(\|x - x^*\|). \tag{2.10}$$

and

$$\overline{U}(x^*, r) \subseteq D, \tag{2.11}$$

where the radius of convergence r is defined in (2.2). Then, sequence $\{x_n\}$ generated for $x_0 \in U(x^*, r) - \{x_n\}$ by method (1.2) is well defined in $U(x^*, r)$, remains in $U(x^*, r)$ for each $n = 0, 1, 2, \dots$ and converges to x^*. Moreover, the following estimates hold

$$\|y_n - x^*\| \le g_1\Big(\|x_n - x^*\|\Big)\|x_n - x^*\| \le \|x_n - x^*\| \le r, \tag{2.12}$$

$$\|z_n - x^*\| \le g_2\Big(\|x_n - x^*\|\Big)\|x_n - x^*\| \le \|x_n - x^*\|, \tag{2.13}$$

and

$$\|x_{n+1} - x^*\| \le g_3\Big(\|x_n - x^*\|\Big)\|x_n - x^*\| \le \|x_n - x^*\|, \tag{2.14}$$

where the functions g_i, for $i = 1, 2, 3$ are defined previously. Furthermore, if there exists $r^* \ge r$ such that

$$\int_0^1 W_0(\theta r^*) d\theta < 1 \tag{2.15}$$

then the point x^* is the only solution of equation $F(x) = 0$ in $D_1 = D \cap \overline{U}(x^*, r^*)$.

Proof. We shall show using mathematical induction that sequence $\{x_n\}$ is well defined, and converges to x^* so that estimates (2.12)-(2.14) hold. Using hypothesis $x_0 \in U(x^*, r) - \{x^*\}$, (2.1), (2.2) and (2.7), we have that

$$\|F'(x^*)^{-1}\Big(F'(x_0) - F'(x^*)\Big)\| \le W_0(\|x_0 - x^*\|) \le W_0(r) < W_0(r_0) = 1. \tag{2.16}$$

It then follows from (2.16) and the Banach lemma on invertible functions [1] that $F'(x_0) \neq 0, y_0$ is well defined by the first sub step of method (1.2) for $n = 0$ and

$$\|F'(x_0)^{-1}F'(x^*)\| \leq \frac{1}{1 - W_0(\|x_0 - x^*\|)}. \tag{2.17}$$

Using (2.2), (2.3) (for $i = 1$), (2.6), (2.8) and (2.17), we have in turn that

$$\begin{aligned}
\|y_0 - x^*\| &= \|x_0 - x^* - F'(x_0)^{-1}F(x_0)\| \\
&\leq \|F'(x_0)^{-1}F'(x^*)\| \, \| \int_0^1 F'(x^*)^{-1}\Big(F'(x^* + \theta(x_0 - x^*)) - F'(x_0)\Big)(x_0 - x^*)\| \\
&\leq \frac{\int_0^1 W\Big((1-\theta)\|x_0 - x^*\|\Big)d\theta\|x_0 - x^*\|}{1 - W_0(\|x_0 - x^*\|)} \\
&= g_1(\|x_0 - x^*\|)\|x_0 - x^*\| \leq \|x_0 - x^*\| < r,
\end{aligned} \tag{2.18}$$

which shows (2.12) for $n = 0$ and $y_0 \in U(x^*, r)$. We can write by (2.6) that

$$F(x_0) = F(x_0) - F(x^*) = \int_0^1 F'\Big(x^* + \theta(x_0 - x^*)\Big)d\theta(x_0 - x^*). \tag{2.19}$$

Notice that $\|x^* + \theta(x_0 - x^*) - x^*\| = \theta\|x_0 - x^*\| < r$, so $x^* + \theta(x_0 - x^*) \in U(x^*, r)$. Then, by (2.9), (2.18) and (2.19), we have that

$$\|F'(x^*)^{-1}F(x_0)\| \leq \int_0^1 V_0(\theta\|x_0 - x^*\|)d\theta\|x_0 - x^*\| \tag{2.20}$$

and

$$\begin{aligned}
\|F'(x^*)^{-1}F(y_0)\| &\leq \int_0^1 V_0\Big(\theta\|y_0 - x^*\|\Big)d\theta\|y_0 - x^*\| \\
&\leq \int_0^1 V_0\Big(\theta g_1(\|x_0 - x^*\|)\Big)\|x_0 - x^*\|d\theta g_1(\|x_0 - x^*\|)\|x_0 - x^*\|.
\end{aligned} \tag{2.21}$$

Next, we must show that $F(x_0) - F(y_0) \neq 0$. In view of (2.4), (2.6), (2.7), and (2.11), we get in turn that

$$\begin{aligned}
&\|\Big(F'(x^*)(x_0-x^*)\Big)^{-1}\Big(F(x_0)-F(x^*)-F'(x^*)(x_0-x^*)-F(y_0)\Big)\| \\
&\leq \|x_0-x^*\|^{-1}\Big[\|\int_0^1 F'(x^*)^{-1}\Big(F'(x^*+\theta(x_0-x^*))-F'(x^*)\Big)(x_0-x^*)d\theta\| \\
&\qquad + \|F'(x^*)^{-1}F(y_0)\|\Big] \\
&\leq \int_0^1 W_0(\theta\|x_0-x^*\|)d\theta + \int_0^1 V_0(\theta g_1(\|x_0-x^*\|))\|x_0-x^*\|d\theta g_1(\|x_0-x^*\|) \\
&= p(\|x_0-x^*\|) \leq p(r) < 1, \qquad (2.22)
\end{aligned}$$

so

$$\|\Big(F(x_0)-F(y_0)\Big)^{-1}F'(x^*)\| \leq \frac{1}{\|x_0-x^*\|\Big(1-p(\|x_0-x^*\|)\Big)}, \qquad (2.23)$$

z_0 is well defined by (2.17), (2.23), $F(x_0) \neq 0$, and the second sub-step of method (1.2) for $n = 0$. Then, we get by (2.2), (2.3) (for $i = 2$), (2.6), (2.17), (2.18), (2.22) and (2.23) that

$$\begin{aligned}
\|z_0-x^*\| &\leq \|x_0-x^*-F'(x_0)^{-1}F(x_0)\| \\
&\quad + \|F'(x_0)^{-1}F'(x^*)\|\;\|\Big(F(x_0)-F(y_0)\Big)^{-1}F'(x^*)\|\|F'(x^*)^{-1}F(y_0)\| \\
&\quad \Big[\|F'(x^*)^{-1}F(x_0)\| + |\alpha-1|\|F'(x^*)^{-1}F(y_0)\|\Big] \\
&\leq g_1(\Psi)\Psi + \frac{\Big[\Psi\int_0^1 V_0(\theta\Psi)d\theta + |\alpha-1|g_1(\Psi)\int_0^1 V_0\Big(\theta g_1(\Psi)\Psi\Big)d\theta\Big]\Psi^2\int_0^1 V_0\Big(\theta g_1(\Psi)\Psi\Big)d\theta}{\Big(1-W_0(\Psi)\Big)\Psi\Big(1-p(\Psi)\Big)} \\
&= g_2(\Psi)\Psi \leq \Psi < r, \qquad (2.24)
\end{aligned}$$

where,

$$\Psi = \|x_0-x^*\|,$$

which shows (2.13) for $n = 0$, and $z_0 \in U(x^*, r)$. We also have as in (2.20) for $z_0 = x_0$ that

$$\|F'(x^*)^{-1}F(z_0)\| \leq \int_0^1 V_0(\theta\|z_0 - x^*\|)d\theta\|z_0 - x^*\|$$
$$\leq \int_0^1 V_0\Big(\theta g_2(\|x_0 - x^*\|)\Big)\|x_0 - x^*\|d\theta\, g_2(\|x_0 - x^*\|)\|x_0 - x^*\|. \tag{2.25}$$

By (2.2),(2.5),(2.7),(2.10),(2.18) and (2.24), we obtain in turn that

$$\begin{aligned}
&\|F'(x^*)^{-1}\Big(F'(x_0) - F'(x^*) - F''(x_0)(z_0 - x_0)\Big)\| \\
&\leq \|F'(x^*)^{-1}\Big(F'(x_0) - F'(x^*)\Big)\| + \|F'(x^*)^{-1}F''(x_0)\|\Big[\|z_0 - x^*\| + \|x_0 - x^*\|\Big] \\
&\leq W_0(\|x_0 - x^*\|) + V(\|x_0 - x^*\|)\Big(1 + g_2(\|x_0 - x^*\|)\Big)\|x_0 - x^*\| \\
&= q(\|x_0 - x^*\|) \leq q(r) < 1,
\end{aligned} \tag{2.26}$$

so

$$\|\Big(F'(x_0) - F''(x_0)(z_0 - x_0)\Big)^{-1}F'(x^*)\| \leq \frac{1}{1 - q(\|x_0 - x^*\|)}, \tag{2.27}$$

and x_1 is well defined by the last sub-step of method (1.2) for $n = 0$. Then, using (2.2), (2.3) (for $i = 3$), (2.25) and (2.27), we get in turn that

$$\begin{aligned}
\|x_1 - x^*\| &\leq \|z_0 - x^*\| + \|\Big(F'(x_0) + F''(x_0)(z_0 - x_0)\Big)^{-1}F'(x^*)\|\ \|F'(x^*)^{-1}F(z_0)\| \\
&\leq \left(1 + \frac{\int_0^1 V_0(\theta(\|z_0 - x^*\|))}{1 - q(\|x_0 - x^*\|)}\right)\|z_0 - x^*\| \\
&\leq \left(1 + \frac{\int_0^1 V_0(\theta g_2\Big(\|x_0 - x^*\|)\|x_0 - x^*\|\Big)d\theta}{1 - q(\|x_0 - x^*\|)}\right)g_2(\|x_0 - x^*\|)\|x_0 - x^*\| \\
&= g_3(\|x_0 - x^*\|)\|x_0 - x^*\| \leq \|x_0 - x^*\| < r,
\end{aligned} \tag{2.28}$$

which shows (2.14) for $n = 0$ and $x_1 \in U(x^*, r)$. By simply replacing x_0, y_0, z_0 by x_k, y_k, z_k, x_{k+1} in the preceding estimates we arrive at (2.12) - (2.14). Then by the estimate

$$\|x_{k+1} - x^*\| \leq \beta \|x_k - x^*\| < r, \tag{2.29}$$

$$\beta = g_3(\|x_0 - x^*\|) \in [0, 1),$$

we deduce that

$$\lim_{k \to \infty} x_k = x^* \text{ and } x_{k+1} \in U(x^*, r).$$

Finally, to show the uniqueness part, let

$$T = \int_0^1 F'\Big(x^* + \theta(y^* - x^*)\Big) d\theta$$

for $y^* \in D_1$ with $F(y^*) = 0$. By (2.7) and (2.15), we get that

$$\|F'(x^*)\Big(T - F'(x^*)\Big)\| \leq \int_0^1 W_0(\theta \|y^* - x^*\|) d\theta \leq \int_0^1 W_0(\theta r^*) d\theta < 1, \tag{2.30}$$

So $T \neq 0$. Using the identity,

$$0 = F(y^*) - F(x^*) = T(y^* - x^*), \tag{2.31}$$

we conclude that $x^* = y^*$.

□

Remark 2.2 (a) It follows from the condition (2.7) that condition (2.9) can be dropped, and be replaced by

$$V(t) = 1 + W_0(t) \text{ or } V(t) = W_0(r_0),$$

since

$$\begin{aligned} \|F'(x^*)^{-1}\Big(F(x) - F(x^*)\Big) + F'(x^*)\| &= 1 + \|F'(x^*)-1\Big(F'(x) - F'(x^*)\Big)\| \\ &\leq 1 + W_0(\|x - x^*\|) = 1 + W_0(r_0) \end{aligned}$$

for $\|x - x^*\| \leq r_0$.

(b) If the function W_0 is strictly increasing, then we choose

$$r_0 = W_0^{-1}(1)$$

instead of $W_0(t) = 1$.

(c) If W_0, W, V are constant functions, then

$$r_1 = \frac{2}{2W_0 + W} \text{ and } r < r_1.$$

Therefore, the radius of convergence r cannot be larger than the radius of convergence r_1 for Newton's method

$$x_{n+1} = x_n - F'(x_n)^{-1} F(x_n).$$

Notice also that the earlier radius of convergence given independently by Rheinboldt [12] and Traud [13] is

$$r_{RT} = \frac{2}{3W_1}$$

and Argyros [1]

$$r_A = \frac{2}{2W_0 + W_1},$$

where W_1 is the Lipschitz constant on Ω. But we have

$$W \leq W_1, W_0 \leq W_1$$

so

$$r_{RT} \leq r_A \leq r_1$$

and

$$\frac{r_{RT}}{r_A} \to \frac{1}{3} \text{ and } \frac{W_0}{W} \to 0.$$

The radius of convergence a_1 used in [12] is smaller than r_{DS} given by Dennis and Schabel,

$$a_1 < r_{DS} = \frac{1}{2W_1} < r_{RT}.$$

However, a_1 cannot be computed using the Lipschitz constants.

(d) We have only used hypotheses on the first-order derivative of F. The order of convergence can be determined by using the following formula for the computational order of convergence (COC) given by

$$\xi = \frac{\ln \|\frac{x_{n+2}-x^*}{x_{n+1}-x^*}\|}{\ln \|\frac{x_{n+1}-x^*}{x_n-x^*}\|}, \text{ for each } n = 1, 2, \ldots$$

or the approximate computational order of convergence (ACOC) [8], given by

$$\xi^* = \frac{\ln \|\frac{x_{n+2}-x_{n+1}}{x_{n+1}-x_n}\|}{\ln \|\frac{x_{n+1}-x_n}{x_n-x_{n-1}}\|}, \text{ for each } n = 1, 2, \ldots$$

that do not require derivatives. Notice also that ξ^* doesn't even require the knowledge of exact root x^*.

(e) The results obtained here can be used for operators F satisfying the autonomous differential equation [1] of the form

$$F'(x) = P(F(x)),$$

where, P is a known continuous operators. Since $F'(x^*) = P(F(x^*)) = P(0)$, we can apply the results without actually knowing the solution of x^*. Let as an example $F(x) = e^x - 1$. Then, we can choose $P(x) = x + 1$.

3. Numerical Examples

Example 3.1 Suppose that the motion of an object is governed by the system of the equation

$$f'(x) - f(x) = 0$$

with $x \in D$ for $f(0) = 0$. Then, the solution is given by $x^* = 0$.
Set $X = Y = \mathbb{R}, D = \overline{U}(0,1)$ and define function f on D by

$$f(x) = e^x - 1.$$

The derivative is given by

$$f'(x) = e^x.$$

Then, we get

$$W_0(t) = (e-1)t,$$

$$W(t) = e^{\frac{1}{e-1}}t,$$

and

$$V_0(t) = V(t) = e^{\frac{1}{e-1}}$$

Choosing $\alpha = 1$ and using equation followed by (2.1), we found that,

$$r_1 \approx 0.382691912232386,$$

$$r_2 \approx 0.197681666516090,$$

and

$$r_3 \approx 0.118698399053728.$$

Thus, using (2.2) we have,

$$r = \min\{r_1, r_2, r_3\} \approx 0.118698399053728.$$

Moreover, Choosing $\alpha = 0.5$ and using equation followed by (2.1), we found that,

$$r_1 \approx 0.382691912232386,$$

$$r_2 \approx 0.193697540794317,$$

and

$$r_3 \approx 0.117495679275249.$$

Thus, using (2.2) we have,

$$r = \min\{r_1, r_2, r_3\} \approx 0.117495679275249.$$

Example 3.2 Returning back to the motivational example, we see that

$$\sup_{x \in D} \frac{|f'(x^*)^{-1}(f'(x) - f'(x^*))|}{|x - x^*|} = 9\ln\left(\frac{3}{2}\right) + \frac{135}{8} - 8,$$

$$\sup_{x,y\in D_0} \frac{|f'(x^*)^{-1}(f'(x)-f'(y))|}{|x-y|} \approx 22.984456878941291,$$

$$V_0(t) = \sup_{x\in D_0} |f'(x^*)^{-1} f'(x)| \approx 1.543816414817695,$$

and

$$V(t) = \sup_{x\in D_0} |f'(x^*)^{-1} f''(x)| \approx 7.661502498177597.$$

Then, using (2.7) and (2.8) we can conclude that,

$$W_0 = \left[9\ln\left(\frac{3}{2}\right) + \frac{135}{8} - 8\right]t,$$

and

$$W \approx 22.984456878941291t.$$

Choosing $\alpha = 1$ and using equation followed by (2.1), we found that,

$$r_1 \approx 0.041638188899749,$$
$$r_2 \approx 0.027347872309846,$$

and

$$r_3 \approx 0.017416938497034.$$

Thus, using (2.2) we have,

$$r = \min\{r_1, r_2, r_3\} \approx 0.017416938497034.$$

Moreover, Choosing $\alpha = 0$ and using equation followed by (2.1), we found that,

$$r_1 \approx 0.041638188899749,$$
$$r_2 \approx 0.027018632548836,$$

and

$$r_3 \approx 0.017309930824510.$$

Thus, using (2.2) we have,

$$r = \min\{r_1, r_2, r_3\} \approx 0.017309930824510.$$

References

[1] Argyros, I.K., *Convergence and Applications of Newton-type Iterations*, Springer Verlag, Berlin, Germany, 2008.

[2] Cordero, A., Ezquerro, J.A., Hernanadez, M.A. and Torregrosa, J.R., On the Convergence of a Fifth-Order Iterative Method in Banach Spaces, *Appl. Math. Comput.* 251 (2015) 396-403.

[3] Dennis, J.E. and Schnabel, R.B., *Numerical Methods for Unconstrained Optimization and Nonlinear Equations*, SIAM, Philadelphia, PA, 1996.

[4] Ezquerro, J.A. and Hernandez, M.A., An Optimization of Chebyshev's Method, *Journal of Complexity 25* (2009) 343-361.

[5] Ezquerro, J.A. and Hernandez, M.A., New Iteration of R-order Four with Reduced Computational Cost, *BIT Numet. Math.* 49 (2010) 325-342.

[6] Hernandez, M.A. and Martinez, E., On the Semi-local Convergence of a Three Steps Newton-type Process under Mild Convergence Conditions, *Numer. Algol.* 70 (2015) 377-392.

[7] Hernandez, M.A. and Romero, N., On the Local Convergence of a Third Order Family of Iterative Processes, *Algorithms* 8 2015 1121-1128.

[8] Kou, J.A., A Third-Order Modification of Newton Method for System of Nonlinear Equations, *Appl. Math. Comput.*, 191 (2007) 117-121.

[9] Maruster, St., Local Convergence of Exquerro-Hernandez Method, Ann. West Univ. Timisoara, *Math. and Comput. Sci.* 54 (1) (2016) 1-9.

[10] Maruster, St., Local Convergence Radius for the Man-type Iteration, West Univ. Timisoara, *Math. and Comput. Sci.* 53 (2015) 109-120.

[11] Maruster, St., Strong Convergence of the Mann Iteration for Demi contractive Mappings, *Appl. Math. Science, HIKARI Idt.*, 9 (42) (2015) 2061-268.

[12] Rheinboldt, W.C., An Adaptive Continuation Process for Solving Systems of Nonlinear Equations, Polish Academy of Science, *Banach Ctr, Publ.* 3 (1979) 129-42.

[13] Traub, J.F., *Iterative Methods for the Solution of Equations*, Chelsea Publishing Company, New York, 1982.

Chapter 3

Ball Convergence of Two Derivative Free Iterative Methods for Solving Equations under Weak Conditions

1. Introduction

Let X and Y be Banach spaces and $D \subset X$ be a non-empty, convex and open set. There is a plethora of problems that can be converted to a nonlinear equation like

$$F(x) = 0, \tag{1.1}$$

where $F : D \to Y$ is continuously differentiable in the sense of Fréchet. The task of finding a locally unique solution x^* in a closed form can be only achieved in special cases. That explains why most researchers and practitioners introduce iterative processes by generating a sequence approximating x^* provided that some convergence conditions are satisfied. Newton's method is without a doubt the most widely used iterative method which converges quadratically, and is defined by

$$x_{n+1} = x_n - F'(x_n)^{-1}F(x_n), \tag{1.2}$$

where $x_0 \in D$ is an initial point. However, there are some short comings when using Newton's method related mainly to the calculation of the inverse of the Fréchet derivative $F'(x)$ at each step. Numerous method have been suggested in the literature. In our chapter, we study the local convergence of methods

$$\begin{aligned} y_n &= x_n - F'\left(\frac{x_{n-1}+y_{n-1}}{2}\right)^{-1} F(x_n) \\ x_{n+1} &= x_n - F'\left(\frac{x_n+y_n}{2}\right)^{-1} F(x_n) \end{aligned} \tag{1.3}$$

and

$$\begin{aligned} y_n &= x_n - \left[y_{n-1}, x_{n-1}; F\right]^{-1} F(x_n) \\ z_n &= x_n - \left[y_n, x_n; F\right]^{-1} F(x_n) \\ x_{n+1} &= z_n - A_n^{-1} F(z_n), \end{aligned} \tag{1.4}$$

where $x_{-1}, y_{-1}, x_0 \in D$ are initial points,

$$A_n = \left[x_n, x_n; F\right] + \left[z_n, x_n, y_n; F\right](z_n - x_n),$$

$\left[\cdot,\cdot;F\right]$ and $\left[\cdot,\cdot,\cdot;F\right]$ are divided differences of order one and two on $D \times D$ and $D \times D \times D$, respectively.

These methods were introduced in [3] in the special cases when $X = Y = \mathbb{R}$. Method (1.3) is of order $1+\sqrt{2}$ and method (1.4) is of order $2+\sqrt{6}$. However, their applicability is limited on $\mathbb{R}$ and the sufficient local convergence conditions involve Taylor expansions and hypotheses on the derivatives up to the order four although such derivates do not appear in these methods.

As an academic example, consider function f defined on the interval $D = [-\frac{1}{2}, \frac{3}{2}]$ by

$$f(x) = \begin{cases} x^3 \ln x^2 + x^5 - x^4, & x \neq 0 \\ 0, & x = 0. \end{cases}$$

Then, we see that $f'''(x)$ is unbounded on D. Hence the results in [3] cannot be used.

The study of convergence of iterative algorithms is usually centered into two categories: semi-local and local convergence analysis. The semi-local convergence is based on the information around an initial point, to obtain conditions ensuring the convergence of these algorithm, while the local convergence is based on the information around a solution to find the estimates of the computed radii of the convergence balls. Local results are important since they provide the degree of difficulty in choosing initial points.

In the present chapter we weaken these condition and the applicability of these methods is extended. Moreover, we avoid Taylor expansion requiring the usage of higher order derivatives but do find the order of convergence using instead the computational and approximate computational order of convergence that do not higher order derivatives.

The rest of the chapter is structured as follows: The second sections contains the local convergence analysis of these method and the third section contains the numerical examples.

2. Local Convergence

We present the local convergence analysis of method (1.2) in this section followed by the local convergence analysis of method (1.3). The convergence analysis of both methods is based on some scalar functions and parameters. Let $W_0 : [0,\infty) \to [0,\infty)$ be a continuous and nondecreasing functions with $W_0(0) = 0$.
Suppose that equation

$$W_0(t) = 1. \tag{2.1}$$

has at least one positive solution denote be r_0 the smallest such solution.
Let also $W, V : [0, r_0) \to [0,\infty)$ be continuous and nondecreasing functions with $W(0) = 0$. Define functions g_1 and h_1 on the interval $[0, r_0)$ by

$$g_1(t) = \frac{\int_0^1 W\Big((1-\theta)t\Big)d\theta}{1 - W_0(t)} + \frac{2W_0(t)\int_0^1 V(\theta t)d\theta}{\Big(1 - W_0(t)\Big)^2}$$

and

$$h_1(t) = g_1(t) - 1.$$

We have that $h_1(0) = -1$ and $h_1(t) \to \infty$ as $t \to r_0^-$. It then follows from the intermediate value theorem that function h_1 has zeros in the interval $(0, r_0)$. Denote by r_1 the smallest such zero. Suppose that equation

$$p(t) = 1. \tag{2.2}$$

has at least one positive solution denote by ρ_0 the smallest such solution. where

$$p(t) = W_0\left(\frac{t + g_1(t)t}{2}\right)$$

Notice that $r_0 \leq \rho_0$, since $0 \leq g_1(t) < 1$ for each $t \in [0, r_0)$. Moreover, define functions g_2 and h_2 on the interval $[0, r_0)$ by

$$g_2(t) = \frac{\int_0^1 W\Big((1-\theta)t\Big)d\theta}{1 - W_0(t)} + \frac{\left[W_0\left(\frac{t+g_1(t)t}{2}\right) + W_0(t)\right]\int_0^1 V(\theta t)d\theta}{\Big(1 - W_0(t)\Big)\Big(1 - p(t)\Big)}$$

and

$$h_2(t) = g_2(t) - 1.$$

We get that $h_2(0) = -1$ and $h_2(t) \to \infty$ as $t \to r_0^-$. Denote by r the smallest zero of function h_2 in the interval $(0, r_0)$. Define a radius of convergence r by,

$$r = \min\{r_1, r_2\} \tag{2.3}$$

Then, for each $t \in [0, r)$

$$0 \leq g_i(t) < 1, \text{ for } i = 1, 2. \tag{2.4}$$

Let $U(u, \rho), \overline{U}(u, \rho)$ stand, respectively for the open and closed balls in X with center $u \in X$ and radius of $\rho > 0$.

Next, we present the local convergence analysis of method (1.2) using the preceding notation.

Theorem 2.1 Let $F : D \subset X \to Y$ be a differential operator in the sense of Fréchet.

Suppose that there exist $x^* \in D$, function $W_0 : [0,\infty) \to [0,\infty)$ continuous and nondecreasing with $W_0(0) = 0$ such that for each $x \in D$

$$F(x^*) = 0, F'(x^*)^{-1} \in L(Y,X) \tag{2.5}$$

and

$$\|F'(x^*)^{-1}\Big(F'(x) - F'(x^*)\Big)\| \leq W_0(\|x - x^*\|). \tag{2.6}$$

Moreover, suppose that there exist functions $W, V : [0,r_0) \to [0,\infty)$ with $W(0) = 0$ such that for each $x, y \in D_0 = D \cap U(x^*, r_0)$

$$\|F'(x^*)^{-1}\Big(F'(x) - F'(y)\Big)\| \leq W(\|x - y\|) \tag{2.7}$$

$$\|F'(x^*)^{-1}F'(x)\| \leq V(\|x - x^*\|) \tag{2.8}$$

and

$$\overline{U}(x^*, r) \subseteq D, \text{ by (2.3)}. \tag{2.9}$$

where r_0 is defined by (2.1) and r followed by (2.3).
Then, the sequence $\{x_n\}$ generated for $x_{-1}, y_{-1}, x_0 \in U(x^*, r) - \{x^*\}$ by method (1.2) is well defined in $U(x^*, r)$ remains in $U(x^*, r)$ for each $n = 0, 1, 2, \ldots$ and converges to x^*. Moreover, the following estimates hold

$$\|y_n - x^*\| \leq g_1(r)\|x_n - x^*\| \leq \|x_n - x^*\| < r, \tag{2.10}$$

and

$$\|x_{n+1} - x^*\| \leq g_2(\|x_n - x^*\|)\|x_n - x^*\| \leq \|x_n - x^*\|, \tag{2.11}$$

where the functions g_i, for $i = 1, 2$ are defined previously. Further more, if there exists $R \geq r$ such that

$$\int_0^1 W_0(\theta R)d\theta < 1, \tag{2.12}$$

then, the limit point x^* is the only solution of equation $F(x) = 0$ in $D_1 = D \cap \overline{U}(x^*, R)$.

Proof. We shall show using mathematical induction on the natural number k that sequence $\{x_n\}$ is well defined, remains in $U(x^*, r)$, converges to x^* and estimates (2.10) and (2.11) are satisfied. Using (2.1), (2.6) and the hypothesis $x_0 \in U(x^*, r) - \{x_n\}$ we have that

$$\|F'(x^*)^{-1}\Big(F'(x_0) - F'(x^*)\Big)\| \leq W_0(\|x_0 - x^*\|) \leq W_0(r) \leq W_0(r_0) < 1. \quad (2.13)$$

It follows from (2.13) and the Banach Lemma on Invertible operators [3] that $F'(x_0)^{-1} \in L(Y, X)$ and

$$\|F'(x_0)^{-1}F'(x^*)\| \leq \frac{1}{1 - W_0(\|x_0 - x^*\|)}. \quad (2.14)$$

Similarly,

$$\begin{aligned} \|F'(x^*)^{-1}\Big(F'\Big(\frac{x_{-1}+y_{-1}}{2}\Big) - F'(x^*)\Big)\| &\leq W_0\Big(\|\frac{x_{-1}+y_{-1}}{2} - x^*\|\Big) \\ &\leq W_0\Big(\frac{\|x_{-1} - x^*\| + \|y_{-1} - x^*\|}{2}\Big) \\ &\leq W_0(r) < 1, \quad (2.15) \end{aligned}$$

so

$$F'\Big(\frac{x_{-1}+y_{-2}}{2}\Big)^{-1} \in L(Y, X),$$

y_0 is well defined by the first sub-step of method (1.2) for $n = 0$ and

$$\begin{aligned} \|F'\Big(\frac{x_{-1}+y_{-1}}{2}\Big)^{-1}F'(x^*)\| &\leq \frac{1}{1 - W_0\Big(\frac{\|x_{-1}-x^*\|+\|y_{-1}-x^*\|}{2}\Big)} \\ &\leq \frac{1}{1 - W_0(r)}. \quad (2.16) \end{aligned}$$

It follows from the first sub-step of method (1.2) for $n = 0$ that we can write the identity

$$\begin{aligned} y_0 - x^* = x_0 - x^* - F'(x_0)^{-1}F(x_0) + F'(x_0)^{-1}\Bigg[\Big(F'\Big(\frac{x_{-1}+y_{-1}}{2}\Big) - F'(x^*)\Big) \\ + \Big(F'(x^*) - F'(x_0)\Big)\Bigg]F'\Big(\frac{x_0+y_0}{2}\Big)^{-1}F(x_0). \quad (2.17) \end{aligned}$$

We can write by (2.5) that

$$F(x_0) = F(x_0) - F(x^*) = \int_0^1 F'\Big(x^* + \theta(x_0 - x^*)\Big) d\theta (x_0 - x^*). \tag{2.18}$$

Notice that $\|x^* + \theta(x_0 - x^*) - x^*\| = \theta\|x_0 - x^*\| < r$, so, $x^* + \theta(x_0 - x^*) \in U(x^*, r)$ for each $\theta \in [0,1]$. Then, we get by (2.8) and (2.18) that

$$\|F'(x^*)^{-1} F(x_0)\| \leq \int_0^1 V(\theta\|x_0 - x^*\|) d\theta \|x_0 - x^*\|. \tag{2.19}$$

In view of (2.4) (for $i = 1$), (2.5), (2.6), (2.7), (2.14), (2.16), (2.17) and (2.19), we get in turn that

$$\begin{aligned}
\|y_0 - x^*\| &\leq \|F'(x_0)^{-1} F'(x^*)\| \, \|\int_0^1 F'(x^*)^{-1} \Big[F'(x^* + \theta(x_0 - x^*)) - F'(x_0)\Big](x_0 - x^*) d\theta\| \\
&+ \|F'(x_0)^{-1} F'(x^*)\| \Big[\|F'(x^*)^{-1} \Big(F'(\frac{x_{-1} + y_{-1}}{2}) - F'(x^*)\Big)\| \\
&+ \|F'(x^*)^{-1} \Big(F'(x^*) - F'(x_0)\Big)\|\Big] \|F'\Big(\frac{x_{-1} + y_{-1}}{2}\Big)^{-1} F'(x^*)\| \, \|F'(x^*)^{-1} F(x_0)\| \\
&\leq \frac{\int_0^1 W\Big((1-\theta)\|x_0 - x^*\|\Big) d\theta \|x_0 - x^*\|}{1 - W_0(\|x_0 - x^*\|)} \\
&+ \frac{\Big[W_0\Big(\frac{\|x_{-1} - x^*\| + \|y_{-1} - x^*\|}{2}\Big) + W_0(\|x_0 - x^*\|)\Big] \int_0^1 V(\theta\|x_0 - x^*\|) d\theta \|x_0 - x^*\|}{\Big(1 - W_0(\|x_0 - x^*\|)\Big)\Big[1 - W_0\Big(\frac{\|x_{-1} - x^*\| + \|y_{-1} - x^*\|}{2}\Big)\Big]} \\
&\leq g_1(r)\|x_0 - x^*\| \leq \|x_0 - x^*\| < r,
\end{aligned} \tag{2.20}$$

which shows (2.10) for $n = 0$ and $y_0 \in U(x^*, r)$. Then, as in (2.15) for x_0, y_0 replacing x_{-1}, y_{-1}, respectively we get that

$$\begin{aligned}
\|F'(x^*)^{-1} \Big(F'(\frac{x_0 + y_0}{2}) - F'(x^*)\Big)\| &\leq W_0\Big(\|\frac{x_0 + y_0}{2} - x^*\|\Big) \\
&\leq W_0\Big(\frac{\|x_0 - x^*\| + \|y_0 - x^*\|}{2}\Big) \\
&\leq W_0\Big(\frac{\|x_0 - x^*\| + g_2(r)\|x_0 - x^*\|}{2}\Big) \\
&\leq W_0(r) < 1,
\end{aligned} \tag{2.21}$$

so $F'\left(\frac{x_0+y_0}{2}\right)^{-1} \in L(Y,X)$, where x_1 is well defined by the second sub-step of method (1.2) for $n=0$, and

$$\|F'\left(\frac{x_0+y_0}{2}\right)^{-1}F'(x^*)\| \leq \frac{1}{1-W_0\left(\frac{\|x_0-x^*\|+\|y_0-x^*\|}{2}\right)} \leq \frac{1}{1-W_0\left(\frac{\|x_0-x^*\|+g(r)\|x_0-x^*\|}{2}\right)}. \tag{2.22}$$

Next, we can write by the second sub-step of method (1.3) that

$$x_1-x^* = x_0-x^*-F'(x_0)^{-1}F(x_0)+F'(x_0)^{-1}\Big[\Big(F'(\frac{x_0+y_0}{2})-F'(x^*)\Big) + F'(x^*)-F'(x_0)\Big]F'\left(\frac{x_0+y_0}{2}\right)^{-1}F(x_0). \tag{2.23}$$

Using (2.4) (for $i=2$), (2.5), (2.6), (2.7), (2.14), (2.16), (2.17), (2.19), (2.20), (2.22) and (2.23) we get in turn that

$$\begin{aligned}
\|x_1-x^*\| &\leq \|F'(x_0)^{-1}F'(x^*)\| \, \|\int_0^1 F'(x^*)^{-1}\Big(F'(x^*+\theta(x_0-x^*)-F'(x_0)\Big)(x_0-x^*)d\theta\| \\
&+\|F'(x_0)^{-1}F'(x^*)\|\Big[\|F'(x^*)^{-1}\Big(F'(\frac{x_0+y_0}{2})-F'(x^*)\Big)\| \\
&+\|F'(x^*)^{-1}(F'(x^*)-F'(x_0))\Big]\| \, \|F'\left(\frac{x_0+y_0}{2}\right)^{-1}F'(x^*)\| \, \|F'(x^*)^{-1}F(x_0)\| \\
&\leq \frac{\int_0^1 W\Big((1-\theta)\|x_0-x^*\|\Big)d\theta\|x_0-x^*\|}{1-W_0(\|x_0-x^*\|)} \\
&+\frac{\Big[W_0\left(\frac{\|x_0-x^*\|+\|y_0-x^*\|}{2}\right)+W_0(\|x_0-x^*\|)\Big]\int_0^1 V(\theta\|x_0-x^*\|)d\theta\|x_0-x^*\|}{\Big(1-W_0(\|x_0-x^*\|)\Big)\Big(1-W_0\big(\frac{\|x_0-x^*\|+\|y_0-x^*\|}{2}\big)\Big)} \\
&\leq g_2(\|x_0-x^*\|)\|x_0-x^*\| \leq \|x_0-x^*\| < r,
\end{aligned} \tag{2.24}$$

so (2.11) holds for $n=0$ and $x_1 \in U(x^*,r)$.

By simply replacing $x_{-1}, y_{-1}, x_0, y_0, x_1$ by $x_{k-1}, y_{k-1}, x_k, y_k, x_{k+1}, y_{k+1}$, respectively in the preceding estimates, we arrive at (2.10) and (2.11).

Then, from the estimates

$$\|x_{k+1}-x^*\| \leq c\|x_k-x^*\| < r, c = g_2(\|x_0-x^*\|) \in [0,1) \tag{2.25}$$

we deduce that

$$\lim_{k\to\infty} x_k = x^* \text{ and } x_{k+1} \in U(x^*,r)$$

. Finally, to show the uniqueness part, let, $y^* \in D_1$ with $F(y^*) = 0$. Let

$$T = \int_0^1 F'\Big(x^* + \theta(y^*-x^*)\Big)d\theta.$$

Using (2.6) and (2.12), we have that

$$\|F'(x^*)^{-1}\Big(T - F'(x^*)\Big)\| \leq \int_0^1 W_0(\theta\|x^*-y^*\|)d\theta \leq \int_0^1 W(\theta R) < 1, \tag{2.26}$$

so $T^{-1} \in L(Y,X)$. By the identity

$$0 = F(y^*) - F(x^*) = T(y^*-x^*),$$

we conclude that $x^* = y^*$.

□

Remark 2.2 (1) In view of (2.6) and the estimate

$$\|F'(x^*)^{-1}F'(x)\| = \|F'(x^*)^{-1}(F'(x)-F'(x^*)) + I\|$$
$$1 + \|F'(x^*)^{-1}(F'(x)-F'(x^*))\| \leq 1 + W_0(\|x-x^*\|)$$

condition (2.8) can be dropped, and V can be replaced by

$$V(t) = 1 + W_0(t).$$

(2) Let $W_0(t) = L_0 t, W(t) = LT, V(t) = M$ for some $L_0 > 0, L > 0$ and $M \geq 1$. In this special case, the results obtained here can be used for operators F satisfying autonomous differential equations [3] of the form

$$F'(x) = P(F(x)),$$

where P is a continuous operator. Then, since $F'(x^*) = P(F(x^*)) = P(0)$, we can apply the results without actually knowing x^*. For example, let $F(x) = e^x - 1$. Then, we can choose: $P(x) = x + 1$.

(3) The radius $r_A = \frac{2}{2L_0+L}$ was shown by us to be the convergence radius of Newton's method [1]

$$x_{n+1} = x_n - F'(x_n)^{-1}F(x_n) \text{ for each } n = 0,1,2\ldots$$

under the conditions $(2.5)-(2.7)$. It follows from the definition of r that the convergence radius r of the method (1.2) cannot be larger than the convergence radius r_A of the second order Newton's method. As already noted in [ ] r_A is at least as large as the convergence ball given by Rheinboldt [3]

$$r_R = \frac{2}{3L}.$$

In particular, for $L_0 < L$ we have that

$$r_R < r$$

and

$$\frac{r_R}{r_A} \to \frac{1}{3} \text{ as } \frac{L_0}{L} \to 0.$$

That is our convergence ball r_A is at most three times larger than Rheinbolt's. The same value for r_R was given by Traub [3].

(4) It is worth noticing that method (1.2) is not changing when we use the conditions of Theorem 2.1 instead of the stronger conditions used in [lolololo]. Moreover, we can compute the computational order of convergence (COC) defined by

$$\xi = \frac{\ln\left(\frac{\|x_{n+1}-x^*\|}{\|x_n-x^*\|}\right)}{\ln\left(\frac{\|x_n-x^*\|}{\|x_{n-1}-x^*\|}\right)}.$$

or the approximate computational order of convergence

$$\xi_1 = \frac{\ln\left(\frac{\|x_{n+1}-x_n\|}{\|x_n-x_{n-1}\|}\right)}{\ln\left(\frac{\|x_n-x_{n-1}\|}{\|x_{n-1}-x_{n-2}\|}\right)}.$$

This way we obtain in practice the order of convergence in a way that avoids the bounds involving estimates using estimates higher than the first Fréchet derivative of operator F.

Let W_0, W, V and r_0 be as in Theorem 2.1. Let also $W_1, W_2 : [0, r_0)^2 \to [0, \infty), V_1 : [0, r_0)^3 \to [0, \infty)$ be continuous, nondecreasing functions with $W_1(0,0) = W_2(0,0) = 0$. Define parameters ρ_0 by

$$\rho_0 = \sup\{t \geq 0 : W_2(t,t) < 1\} \tag{2.27}$$

and set

$$\bar{r}_0 = \min\{x_0, \rho_0\}. \tag{2.28}$$

Define functions g_1 and h_1 on the interval $[0, \bar{r}_0)$ by

$$g_1(t) = \frac{\int_0^1 W(1-\theta)t)d\theta}{1 - W_0(t)} + \frac{W_1(2t,2t)\int_0^1 V(\theta t)d\theta}{(1 - W_0(t))(1 - W_2(t,t))}$$

and

$$h_1(t) = g_1(t) - 1.$$

We have that $h_1(0) = -1$ and $h_1(t) \to \infty$ as $t \to \bar{r}_0$. Denote by r_1 the smallest zero of function h_1 on the interval $(0, \bar{r}_0)$.

Moreover, define functions g_2 and h_2 on the interval $[0, \bar{r}_0)$ by

$$g_2(t) = \frac{\int_0^1 W((1-\theta)t)d\theta}{1 - W_0(t)} + \frac{W_1(2t,2t)\int_0^1 V(\theta t)d\theta}{\Big(1 - W_0(t)\Big)\Big(1 - W_2\big(g_1(t)t, t\big)\Big)}$$

and

$$h_2(t) = g_2(t) - 1.$$

We get that $h_2(0) = -1$ and $h_2(t) \to \infty$ as $t \to \bar{r}_0^-$. Denote by r the smallest zero of function h_2 on the interval $(0, \bar{r}_0)$.
Notice that $h_2(r_1) = 0$, since $g_1(r_1) = 1$, so $r \leq r_1$. Furthermore, define function

$$p(t) = V_1\Big(g_2(t)t, t, g_1(t)t\Big)(1 + g_2(t))t + W_2\big(g_2(t)t, t\big)$$

parameter ρ_1 by

$$\rho_1 = \sup\{t \in [0, r_0) : p(t) < 1\}.$$

and set

$$\rho_2 = \min\{\bar{r}_0, \rho_1\}.$$

Finally, defined function g_3 and h_3 on the interval $[0, \rho_2)$ by

$$g_3(t) = \left(1 + \frac{\int_0^1 V\left(\theta g_2(t)t\right)d\theta}{1 - p(t)}\right)g_2(t)$$

and

$$h_3(t) = g_3(t) - 1.$$

we have that $h_3(0) = -1$ and $h_3(t) \to \infty$ as $t \to \rho_2^-$. Denote by r_3 the smallest zero of functions h_3 on the interval $(0, \rho_2)$. set

$$r = \min\{r, r_3\}. \tag{2.29}$$

Then, we have that

$$0 \leq g_i(t) < 1, \text{ for } i = 1,2,3, \tag{2.30}$$

$$W_0(t) < 1,$$

$$0 \leq W_2(t,t) < 1, \tag{2.31}$$

and

$$0 \leq p(t) < 1. \tag{2.32}$$

Next, we present the local convergence analysis of method (1.3) in an analogous to Theorem 2.1 way.

Theorem 2.3 Let $F : D \subset X \to Y$ be a Fréchet-differentiable operator and let $[\cdot, \cdot; F], [\cdot, \cdot, \cdot; F]$ be divided differences of order one and two for F on D^2 and D^3 respectively.

Suppose that there exist $x^* \in D$, and a function $W_0 : [0, \infty) \to [0, \infty)$ continuous and nondecreasing with $W_0(0) = 0$ such that for each $x \in D$.

$$F(x^*) = 0, F'(x^*)^{-1} \in L(Y, X)$$

and

$$\|F'(x^*)^{-1}\Big(F'(x)-F'(x^*)\Big)\| \leq W_0(\|x-x^*\|).$$

Moreover, there exist functions $W, V : [0, r_0) \to [0, \infty), W_1, W_2 : [0, r_0)^2 \to [0, \infty)$ and $V_1 : [0, +r_0)^3 \to [0, \infty)$ continuous and nondecreasing with $W(0) = r(0) = W_1(0, 0) = W_2(0, 0) = 0$ such that for each $x, y, z \in D_0 = D \cap U(x^*, r_0)$

$$\|F'(x^*)^{-1}\Big(F'(x)-F'(y)\Big)\| \leq W(\|x-y\|),$$

$$\|F'(x^*)^{-1}F'(x)\| \leq V(\|x-x^*\|),$$

$$\|F'(x^*)^{-1}\Big([x,y;F]-F'(z)\Big)\| \leq W_1\big(\|x-z\|, \|y-z\|\big), \tag{2.33}$$

$$\|F'(x^*)^{-1}\Big([x,y;F]-F'(x^*)\Big)\| \leq W_1\big(\|x-x^*\|, \|y-x^*\|\big), \tag{2.34}$$

$$\|F'(x^*)^{-1}[x,y,z;F]\| \leq V_1\big(\|x-x^*\|, \|y-x^*\|, \|z-x^*\|\big), \tag{2.35}$$

and

$$\overline{U}(x^*, r) \subseteq D,$$

where r_0, r are defined by (2.1) and (2.29), respectively. Then ,the sequence $\{x_n\}$ generated for $x_{-1}, y_{-1}, x_0 \in U(x^*, r) - \{x^*\}$ by method (1.3) is well defined in $U(x^*, r)$, remain in $U(x^*, r)$ for each $n = 0, 1, 2, \ldots$ and converges to x^*. Moreover, the following estimates hold

$$\|y_n - x^*\| \leq g_1(r)\|x_n - x^*\| \leq \|x_n - x^*\| < r, \tag{2.36}$$

$$\|z_n - x^*\| \leq g_2(\|x_n - x^*\|)\|x_n - x^*\| \leq \|x_n - x^*\|, \tag{2.37}$$

$$\|x_{n+1} - x^*\| \leq g_3(\|x_n - x^*\|)\|x_n - x^*\| \leq \|x_n - x^*\|, \tag{2.38}$$

where, the functions g_i, for $i = 1, 2, 3$ are defined previously. Furthermore, if there exists $R_1 \geq r$ such that

$$\int_0^1 W_0(\theta R_1)d\theta < 1, \tag{2.39}$$

then the limits point x^* is the only solution of equation $F(x) = 0$ in $D_2 = D \cap \overline{U}(x^*, r)$.

Proof. We follow the proof of Theorem 2.1 and also use some of the estimates obtained in that proof. Using (2.31) and (2.38), we have that

$$\|F'(x^*)^{-1}\Big([y_{-1},x_{-1};F]-F'(x^*)\Big)\| \le W_2(\|y_{-1}-x^*\|,\|x_{-1}-x^*\|) \le W_2(r,r) < 1 \tag{2.40}$$

so

$$\|[y_{-1},x_{-1};F]^{-1}F'(x^*)\| \le \frac{1}{1-W_2(\|y_{-1}-x^*\|,\|x_{-1}-x^*\|)} \tag{2.41}$$

By the first sub-step of method (1.3) for $n=0$, we have the identity

$$y_0-x^* = x_0-x^*-F'(x_0)^{-1}F(x_0)+F'(x_0)^{-1}\Big([y_{-1},x_{-1};F-F'(x_0)]\Big)F'(x_0)^{-1}F(x_0) \tag{2.42}$$

Then, by (2.5), (2.6), (2.7), (2.8), (2.29), (2.30) (for $i=1$), (2.33), (2.41) and (2.42), we have in turn that

$$\begin{aligned}
\|y_0-x^*\| &\le \|x_0-x^*-F'(x_0)^{-1}F(x_0)\|+\|F'(x_0)^{-1}F'(x^*)\| \\
&\quad \|F'(x^*)^{-1}\Big([y_{-1},x_{-1};F]-F'(x_0)\Big)\|\ \|[y_{-1},x_{-1};F]^{-1}F'(x^*)\|\ \|F'(x^*)^{-1}F(x_0)\| \\
&\le \frac{\int_0^1 W\Big((1-\theta)\|x_0-x^*\|\Big)d\theta\|x_0-x^*\|}{1-W_0(\|x_0-x^*\|)} \\
&\quad + \frac{W_1(\|y_{-1}-x_0\|,\|x_{-1}-x_0\|)\int_0^1 V\Big(\theta\|x_0-x^*\|\Big)d\theta\|x_0-x^*\|}{\Big(1-W_0(\|x_0-x^*\|)\Big)\Big(1-W_2(\|y_{-1}-x^*\|,\|x_{-1}-x^*\|)\Big)} \\
&\le g_1(r)\|x_0-x^*\|,
\end{aligned} \tag{2.43}$$

which shows (2.36) for $n=0$ and $y_0 \in U(x^*,r)$.
Similarly, using the identity

$$\begin{aligned}
z_0-x^* &= x_0-x^*-F'(x_0)^{-1}F(x_0)+\Big[\Big(F'(x_0)^{-1}-F'(x^*)\Big) \\
&\quad \Big(F'(x^*)^{-1}[y_0,x_0;F]^{-1}-F'(x^*)\Big)\Big(F'(x^*)^{-1}F(x_0)\Big)\Big],
\end{aligned} \tag{2.44}$$

we obtain

$$\|z_0 - x^*\| \leq \frac{\int_0^1 W\big((1-\theta)\|x_0 - x^*\|\big)d\theta\|x_0 - x^*\|}{1 - W_0(\|x_0 - x^*\|)}$$
$$+ \frac{W_1(\|y_0 - x_0\|, \|x_0 - x_0\|)\int_0^1 V\Big(\theta\|x_0 - x^*\|\Big)d\theta\|x_0 - x^*\|}{\Big(1 - W_0(\|x_0 - x^*\|)\Big)\Big(1 - W_2(\|y_0 - x^*\|, \|x_0 - x^*\|)\Big)}$$
$$\leq g_2(\|x_0 - x^*\|)\|x_0 - x^*\| \leq \|x_0 - x^*\| < r, \tag{2.45}$$

which shows (2.37) for $n = 0$ and $z_0 \in U(x^*, r)$. We must show that $A_0^{-1} \in L(Y, X)$.
In view of (2.32), (2.34), (2.35), (2.43) and (2.45), we get that

$$\|F'(x^*)^{-1}\Big(A_0 - F'(x^*)\Big)\| \leq \|F'(x^*)^{-1}\Big([z_0, x_0; F] - F'(x^*)\Big)\| + \|F'(x^*)^{-1}[z_0, x_0, y_0]\|$$
$$\leq W_2(\|z_0 - x^*\|, \|x_0 - x^*\|) + V_1\Big(\|z_n - x^*\|, \|x_n - x^*\|, \|y_n - x^*\|\Big)\Big(\|z_n - x^*\| + \|x_n - x^*\|\Big)$$
$$\leq W_2\Big(g_2(\|x_0 - x^*\|)\|x_0 - x^*\|, \|x_0 - x^*\|, g_1(r)\|x_0 - x^*\|\Big)$$
$$+ V_1\Big(\|z_n - x^*\|, \|x_n - x^*\|, \|y_n - x^*\|\Big)\Big(\|z_n - x^*\| + \|x_n - x^*\|\Big)$$
$$\leq W_2\Big(g_2(\|x_0 - x^*\|)\|x_0 - x^*\|, \|x_0 - x^*\|, g_1(r)\|x_0 - x^*\|\Big)$$
$$+ V_1\Big(g_2(\|x_0 - x^*\|)\|x_0 - x^*\|, \|x_0 - x^*\|, g_1(r)\|x_0 - x^*\|\Big)\Big(1 + g_2(\|x_0 - x^*\|)\Big)\|x_0 - x^*\|$$
$$= p(\|x_0 - x^*\|) \leq p(r) < 1, \tag{2.46}$$

so

$$\|A_0^{-1}F'(x^*)\| \leq \frac{1}{1 - p(\|x_0 - x^*\|)}. \tag{2.47}$$

Next, by the last sub-step of method (1.3) for $n = 0$, (2.19) (for $x_0 = z_0$), (2.30) (for $i = 3$), (2.45) and (2.47), we obtain in turn that

$$\|x_1 - x^*\| \leq \|z_0 - x^*\| + \|A_0^{-1}F'(x^*)\|\,\|F'(x^*)^{-1}F(z_0)\|\left(1 + \frac{\int_0^1 V(\theta\|z_0 - x^*\|)d\theta}{1 - p(\|x_0 - x^*\|)}\right)\|z_0 - x^*\|$$
$$\leq \left(1 + \frac{\int_0^1 V\Big(\theta g_2(\|x_0 - x^*\|)\|x_0 - x^*\|\Big)d\theta}{1 - p(\|x_0 - x^*\|)}\right)g_2(\|x_0 - x^*\|)\|x_0 - x^*\|$$
$$= g_3(\|x_0 - x^*\|)\|x_0 - x^*\| \leq \|x_0 - x^*\| < r, \tag{2.48}$$

which shows (2.38) for $n = 0$ and $x_1 \in U(x^*, r)$. By simply replacing $x_{-1}, y_{-1}, x_0, y_0, z_0, x_1$ by $x_{k-1}, y_{k-1}, x_k, y_k, z_k, x_{k+1}$ in the preceding estimates we arrive at (2.36) - (2.38). Then, from the estimates (2.41)

$$\|x_{k+1} - x^*\| \leq d\|x_k - x^*\| < r,\ d = g_3\big(\|x_0 - x^*\|\big) \in [0, 1),$$

we deduce that

$$\lim_{k \to \infty} x_k = x^* \text{ and } x_{k+1} \in U(x^*, r).$$

Finally, the uniqueness part has been shown in Theorem 2.1.

□

Remark 2.4 Condition (2.33) can be dropped from Theorem 2.3. In this case functions g_1, g_2, g_3 can be defined by

$$\overline{g}_1(t) = \frac{\int_0^1 W\Big((1-\theta)t\Big)d\theta}{1 - W_0(t)} + \frac{\Big(W_2(t,t) + W_0(t)\Big)\int_0^1 V(\theta t)d\theta}{\Big(1 - W_0(t)\Big)\Big(1 - W_2(t,t)\Big)},$$

$$\overline{g}_2(t) = \frac{\int_0^1 W\Big((1-\theta)t\Big)d\theta}{1 - W_0(t)} + \frac{\Big(W_2(t,t) + W_0(t)\Big)\int_0^1 V(\theta t)d\theta}{\Big(1 - W_0(t)\Big)\Big(1 - W_2\big(g_1(t)t, t\big)\Big)}$$

and $\overline{g}_3$ is as previously defined but g_1, g_2 are replaced by $\overline{g}_1, \overline{g}_2$, respectively. These changes are due to the fact that using just (2.34) instead of (2.33), we get that

$$\begin{aligned}
\|F'(x^*)^{-1}\Big([y_{k-1}, x_{k-1}; F] - F(x_k)\Big)\| &\leq \|F'(x^*)^{-1}\Big([y_{k-1}, x_{k-1}; F] - F(x^*)\Big)\| \\
&\quad + \|F'(x^*)^{-1}\Big(F'(x_k) - F'(x^*)\Big)\| \\
&\leq W_2\Big(\|y_{k-1} - x^*\|, \|x_{k-1} - x^*\|\Big) \\
&\leq W_2\Big(\|x_{k-1} - x^*\|, \|x_{k-1} - x^*\|\Big) \\
&\leq W_2\big(r, r\big) + W_0(r).
\end{aligned}$$

Then, the conclusions of Theorem 2.3 hold in this setting.

Theorem 2.5 Suppose that the hypotheses of Theorem 2.3 hold except (2.33). Then, then conclusions for sequence $\{x_n\}$ hold but with function $\overline{g}_i$, replacing g_i, for $i = 1,2,3$, respectively in estimates (2.36)-(2.38) and in the computation of the radius $\overline{r}$ corresponding to r.

3. Numerical Examples

Example 3.1 Suppose that the motion of an object is governed by the system of the equation

$$f'(x) - f(x) = 0$$

with $x \in D$ for $f(0) = 0$. Then, the solution is given by $x^* = 0$.
Set $X = Y = \mathbb{R}, D = \overline{U}(0,1)$ and define function f on D by

$$f(x) = e^x - 1.$$

The derivative is given by

$$f'(x) = e^x.$$

Then, using equation (2.6), (2.7), and (2.8), we get,

$$W_0(t) = (e-1)t,$$

$$W(t) = e^{\frac{1}{e-1}}t,$$

and

$$V(t) = e^{\frac{1}{e-1}}$$

Using equations followed by (2.1) and (2.2), we found that,

$$r_1 = 0.099662690628594,$$

and

$$r_2 = 0.099662690628594.$$

Thus, using (2.3) we have,

$$r = \min\{r_1, r_2\} = 0.099662690628594.$$

Example 3.2 Returning back to the motivational example, we see that

$$\sup_{x\in D} \frac{|f'(x^*)^{-1}(f'(x)-f'(x^*))|}{|x-x^*|} = 9\ln\left(\frac{3}{2}\right)+\frac{135}{8}-8,$$

$$\sup_{x,y\in D_0} \frac{|f'(x^*)^{-1}(f'(x)-f'(y))|}{|x-y|} \approx 22.984456878941291,$$

and

$$V(t) = \sup_{x\in D_0} |f'(x^*)^{-1}f'(x)| \approx 1.543816414817695.$$

Then, using (2.6) and (2.7) we can conclude that,

$$W_0 = \left[9\ln\left(\frac{3}{2}\right)+\frac{135}{8}-8\right]t,$$

and

$$W \approx 22.984456878941291t.$$

Using equation followed by (2.1) to (2.3), we found that,

$$r_1 \approx 0.014089938337093,$$

and

$$r_2 \approx 0.014089938337093.$$

Thus, using (2.3) we have,

$$r = \min\{r_1, r_2\} \approx 0.014089938337093.$$

References

[1] Argyros, I.K., *Convergence and Applications of Newton-type Iterations*, Springer Verlag, Berlin, Germany, 2008.

[2] Argyros, I.K. and Magreñán, A.A., *A Contemporary Study of Iterative Methods*, Academic Press, Elsevier, San Diego, CA, USA, 2017.

[3] Kantorovich, L.V. and Akilov, G.P., *Functional Analysis in Normed Spaces*, Pergamon Press, New York, 1964.

Chapter 4

Quasi-Newtonian Iterative Processes

1. Introduction

Let X, Y be the Banach spaces and D be an nonempty, convex and open subset of X. Numerous problems in diverse disciplines can be written in the form

$$F(x) = 0, \tag{1.1}$$

where $F : D \subset X \to Y$ is a continuously differentiable operator in the sense of Fréchet. The task of finding a locally unique solution x^* of equation (1.1) in closed form is difficult in general. That is why researchers and practitioners resort to iterative methods generating a sequence converging to x^*.

The most popular iterative process is without a doubt Newton's method defined for each $n = 0, 1, 2, \ldots$ by

$$x_{n+1} = G(x_n), \tag{1.2}$$

where

$$G(x) = x - F'(x)^{-1}F(x).$$

A possible approximation of the fixed point $x = G(x)$ is the Mann's Iteration [5] defined for each $n = 0, 1, 2, \ldots$ by

$$x_{n+1} = (1 - \alpha_n)x_n + \alpha_n G(x_n), \tag{1.3}$$

where

$$\sum \alpha_n = \infty, 0 < \alpha_n < 1, \text{ and } \lim_{n\longrightarrow\infty} \alpha_n = 0.$$

By substituting (1.2) into (1.3), we obtain a new iteration Quasi-Newton Iterative process (QNIP) defined for each $n = 0, 1, 2, \ldots$ by

$$x_{n+1} = x_n - \alpha_n F'(x_n)^{-1} F(x_n). \tag{1.4}$$

The study of convergence of iterative algorithm is usually centered into two categories: semi-local and local convergence analysis. The semi-local convergence is based on the information around an initial point, to obtain conditions ensuring the convergence of these algorithms while the local convergence is based on the information around a solution to find estimates of the computed radii of the convergences balls. Local results are important since they provide the degree of difficulty in choosing initial points.

In this chapter, we present the local as well as the semi-local convergence analysis of method (1.3) under weaker condition than in earlier works using Lipschitz or average Lipschitz conditions. The rest of the chapter is structured as follows: Section 2 and Section 3 contains the local and semi-local analysis of method (1.3), respectively. While, the section 4 contains the numerical example.

2. Local Convergence

The local convergence analysis of QNIP is based on some scalar functions and parameters.
Let $W_0 : [0, \infty) \to [0, \infty)$ be a continuous and nondecreasing function with $W_0(0) = 0$.
Suppose that equation

$$W_0(t) = 1, \tag{2.1}$$

has at-least one positive solution. Denoted by r_0 the smallest such solution. Let also α_n be a sequence of real numbers, $W : [0, r_0) \to [0, \infty)$ and $V : [0, r_0) \to [0, \infty)$ be continuous and nondecreasing functions with W(0) = 0. Suppose that

$$1 - \frac{1}{V(0)} < \alpha_n < 1 + \frac{1}{V(0)}. \tag{2.2}$$

Moreover, define functions g_n and h_n on the interval $[0, r_0)$ by

$$g_n(t) = \frac{\int_0^1 W((1-\theta)t)d\theta + |1-\alpha_n| \int_0^1 V(\theta t)d\theta}{1 - W_0(t)}$$

and

$$h_n(t) = g_n(t) - 1.$$

We have by (2.1) and (2.2) that $h_n(0) = |1-\alpha_n|V(0) - 1 < 0$ and $h_n(t) \to \infty$ as $t \to r_0^-$. Hence, functions h_n have zeros in the interval $(0, r_0)$ by the intermediate value theorem. Denote by r_n the smallest zeros of function h_n on the interval $(0, r_0)$, respectively.

Furthermore, suppose that

$$r = \inf_{n \geq 0} r_n > 0. \tag{2.3}$$

Then, we have that for each $t \in [0, r)$

$$0 \leq W_0(t) < 1, \tag{2.4}$$

and

$$0 \leq g_n(t) < 1 \text{ for each } n = 0, 1, 2, \ldots \tag{2.5}$$

Let $U(w, \delta), \overline{U}(w, \delta)$ stand, respectively for the open and closed balls in D with center $w \in D$ and of radius $\delta > 0$.

Next, we give the local convergence analysis of the QNIP using the preceding notation.

Theorem 2.1 Let $F : D \subseteq X \to Y$ be a continuously Fréchet-differentiable operator. Suppose: there exists $x^* \in D$, function $W_0 : [0, \infty) \to [0, \infty)$ continuous and nondecreasing with $W_0(0) = 0$ such that for each $x \in D, F(x^*) = 0$, $F'(x^*)$ is invertible,

$$\|F'(x^*)^{-1}\left(F'(x) - F'(x^*)\right)\| \leq W_0(\|x - x^*\|) \tag{2.6}$$

Let $D_0 = D \cap U(x^*, r_0)$

There exist functions $W : [0, r_0) \to [0, \infty)$, $V : [0, r_0) \to [0, \infty)$ continuous and nondecreasing functions with W(0) = 0 such that for each $x, y \in D_0$

$$\|F'(x^*)^{-1}\Big(F'(x)-F'(y)\Big)\| \leq W(\|x-y\|), \tag{2.7}$$

$$\|F'(x^*)^{-1}F'(x)\| \leq V(\|x-x^*\|). \tag{2.8}$$

There exists a sequence of real numbers $\{\alpha_n\}$ satisfying (2.2).
The limit point r satisfies (2.3) and

$$\overline{U}(x^*, r) \subseteq D. \tag{2.9}$$

Then, iteration $\{x_n\}$ generated for $x_0 \in U(x^*, r) - \{x^*\}$ by QNIP is well defined $U(x^*, r)$ remains in $U(x^*, r)$ for each $n = 0, 1, \ldots$ and converges to x^*. Moreover, the following estimates hold

$$\|x_{n+1}-x^*\| \leq g_n(\|x_n-x^*\|)\|x_n-x^*\| \leq \|x_n-x^*\| < r, \tag{2.10}$$

where the g_n functions are defined previously and the convergence radius r is given by (2.3). Furthermore, if there exists $r^* \geq r$ such that

$$\int_0^1 W_0(\theta r^*)d\theta < 1, \tag{2.11}$$

then, the limit point x^* is the only solution of equation $F(x) = 0$ in $D_1 = D \cap \overline{U}(x^*, r^*)$.

Proof. The proof uses mathematical induction on k. Using (2.1), (2.3), (2.4) and (2.6) and the hypotheses $x_0 \in U(x^*, r^*) - \{x^*\}$ we get in turn that

$$\|F'(x^*)^{-1}\Big(F'(x_0)-F'(x^*)\Big)\| \leq \|W_0(\|x_0-x^*\|)\| \leq W_0(r) < 1. \tag{2.12}$$

It follows from (2.12) and the Banach space lemma on invertible operators [1,2,4] that $F'(x_0)$ is invertible, and

$$\|F'(x_0)^{-1}F'(x^*)\| \leq \frac{1}{1-W_0(\|x_0-x^*\|)}. \tag{2.13}$$

Hence, x_1 is well defined by QNIP for $n = 0$. We can write

$$F(x_0) = F(x_0) - F(x^*) = \int_0^1 F'\Big(x^* + \theta(x_0 - x^*)\Big)d\theta(x_0 - x^*). \tag{2.14}$$

Notice that $\|x^* + \theta(x_0 - x^*)\| = \theta\|x_0 - x^*\| < r$, so $x^* + \theta(x_0 - x^*) \in U(x^*, r)$ for each $\theta \in [0,1]$. By (2.8) and (2.14) we have that

$$\|F'(x^*)^{-1}F(x_0)\| \leq \int_0^1 V\Big(\theta\|x_0 - x^*\|\Big)\|x_0 - x^*\|d\theta. \tag{2.15}$$

In view of QNIP was can have that

$$x_1 = x_0 - x^* - F'(x_0)^{-1}F(x_0) + (1-\alpha)F'(x_0)^{-1}F(x_0). \tag{2.16}$$

Using (2.1), (2.3), (2.5), (2.7), (2.13), (2.15) and (2.16), we obtain in turn that

$$\begin{aligned}
\|x_1 - x^*\| &\leq \|F'(x_0)^{-1}F'(x^*)\| \|\int_0^1 F'(x^*)^{-1}\Big(F'(x^* + \theta(x_0 - x^*)) - F'(x_0)\Big)(x_0 - x^*)d\theta\| \\
&\quad + |1-\alpha_0| \|F'(x_0)^{-1}F'(x^*)\| \|F'(x^*)^{-1}F(x_0)\| \\
&\leq \frac{\Big[\int_0^1 W(1-\theta)\|x_0 - x^*\|d\theta + |1-\alpha_0|\int_0^1 V\Big(\theta\|x_0 - x^*\|\Big)d\theta\Big]\|x_0 - x^*\|}{1 - W_0(\|x_0 - x^*\|)} \\
&= g_1(\|x_0 - x^*\|)\|x_0 - x^*\| \leq \|x_0 - x^*\| < r,
\end{aligned} \tag{2.17}$$

which shows (2.10) for $n = 0$ and $x_1 \in U(x^*, r)$.

The induction is completed, if we simply replace x_0, x_1 by x_k, x_{k+1} in the preceding estimates. Then, in view of estimates

$$\|x_{k+1} - x^*\| \leq c\|x_k - x^*\| < r, \tag{2.18}$$

where $c = g_1(\|x_0 - x^*\|) \in [0,1)$, we deduce that

$$\lim_{k \to \infty} x_k = x^* \text{ and } x_{k+1} \in U(x^*, r).$$

Finally, to show the uniqueness part, let $y^* \in D_1$ with $F(y^*) = 0$. Define the linear operator

$$Q = \int_0^1 F'\Big(x^* + \theta(y^* - x^*)\Big)d\theta.$$

Then, by (2.6) and (2.11), we get in turn that

$$\|F'(x^*)^{-1}\Big(Q - F'(x^*)\Big)\| \leq \int_0^1 W_0\Big(\theta\|x^* - y^*\|\Big)d\theta \leq \int_0^1 W_0(\theta r^*)d\theta < 1, \quad (2.19)$$

so operator Q is invertible. Using the operator identity.

$$0 = F(y^*) - F(x^*) = Q(y^* - x^*), \quad (2.20)$$

we can conclude that $x^* = y^*$. □

3. Semi-Local Convergence

We shall use the following auxiliary result for the semi-local convergence analysis of the QNIP that follows:

Lemma 3.1 [3] Let $\{x_n\}, \{\beta_n\}, \{\gamma_n\}$ be nonnegative sequences satisfying

$$\beta_{n+1} \leq (1 - \alpha_n) + \alpha_n \gamma_n \text{ for all } n = 0, 1, 2, \ldots, \quad (3.1)$$

$$\sum_{n=0}^{\infty} \alpha_n = \infty, \lim_{n \to +\infty} \alpha_n = 0, \alpha_n \in (0,1), \text{and } \lim_{n \to +\infty} \gamma_n = 0. \quad (3.2)$$

Then, the following item holds

$$\lim_{n \to +\infty} \beta_n = 0. \quad (3.3)$$

We shall use the conditions (A):

(A_1) $F : D \subseteq X \to Y$ is a continuously Fréchet-differentiable operator.

(A_2) There exists function $u_0 : [0,\infty) \to [0,\infty)$ continuous, nondecreasing with $u_0(0) = 0$ and $x_0 \in D$ such that for each $x \in D, F'(x_0)$ is invertible,

$$\|F'(x_0)^{-1}\Big(F'(x) - F'(x_0)\Big)\| \leq u_0(\|x - x_0\|).$$

Suppose that equation

$$u_0(t) = 1$$

has at least one positive solution. Denote by ρ_0 the smallest solution such solution.

Define also set D_2 by

$$D_2 = D \cap U(x_0, \rho_0).$$

(A_3) There exists function $u : [0,\rho_0] \to [0,\infty)$ continuous, nondecreasing with $u(0) = 0$ such that for each $x, y \in D_2$

$$\|F'(x_0)^{-1}\Big(F'(x) - F'(y)\Big)\| \leq u(\|x - y\|).$$

(A_4) $\|F'(x_0)^{-1}F(x_0)\| \leq \delta$ for some $\delta > 0$, and

$$\mu_0 := \int_0^1 u\Big(\frac{\theta\alpha_0\delta}{1 - u_0(\rho_0)}\Big) d\theta \frac{1}{1 - u_0(\rho_0)} \leq 1.$$

(A_5) $\overline{U}(x_0, \rho) \subseteq D$,

where,

$$\rho = \frac{\lambda\delta}{\Big(1 - u_0(\rho_0)\Big)(1 - \lambda)} + \alpha_0\|F'(x_0)^{-1}F(x_0)\|$$

for some $\lambda \in (0,1)$.

(A_6) For $\rho^* \geq \rho$

$$\int_0^1 u_0\Big(\theta\rho + (1-\theta)\rho^*\Big) d\theta < 1.$$

Set $D_3 = D \cap \overline{U}(x_0, \rho^*)$.

(A_7) $\rho \leq \rho_0$.

(A_8) There exists sequence $\{x_n\}$ such that

$$\sum \alpha_n = \infty, \lim_{n\to\infty} \alpha_n = 0, \alpha_n \in (0,1) \text{ for each } n = 0,1,2,\ldots$$

Theorem 3.2 Suppose that conditions (A) hold. Then, sequence $\{x_n\}$ generated for $x_0 \in D$ by QNIP is well defined in $U(x_0,\delta)$, remains in $U(x_0,\delta)$ for each $n = 0,1,2,\ldots$ and converges to a unique solution x^* of equation $F(x) = 0$ in D_3. Moreover, the following estimates hold

$$\|x_n - x^*\| \leq \frac{\xi\delta\lambda^n}{1-\lambda} \tag{3.4}$$

Proof. Let $x \in U(x_0,\rho)$. Using $(A_2), (A_7)$ and the definition δ_0 we have in turn that

$$\|F'(x_0)^{-1}\Big(F'(x) - F'(x_0)\Big)\| \leq u_0(\|x - x_0\|) \leq u(\rho_0) \leq u_0(r_0) < 1$$

Hence, $F'(x)$ is invertible and

$$\|F'(x)F'(x_0)\| \leq \frac{1}{1 - u_0(\|x - x_0\|)}. \tag{3.5}$$

The iterate x_1 is well defined by QNIP for $b = 0$ and (A_2). Set $\beta_k := \|F'(x_0)^{-1}F(x_k)\|$. We get from QNIP and (3.5) for $x = x_k$ that

$$\begin{aligned}
\|x_{k+1} - x_k\| &= \|\alpha_k F'(x_k)^{-1}F(x_k)\| \\
&\leq \alpha_k \|F'(x_k)^{-1}F'(x_0)\| \, \|F'(x_0)^{-1}F(x_k)\| \\
&\leq \frac{\alpha_k \|F'(x_0)^{-1}F(x_k)\|}{1 - u_0(\|x_k - x_0\|)} = \frac{\alpha_k\beta_k}{1 - u_0(\|x_k - x_0\|)} \\
&= \alpha_k\beta_k\xi_k \leq \alpha_k\beta_k\xi,
\end{aligned} \tag{3.6}$$

where,

$$\xi_k := \frac{1}{1 - u_0(\|x_k - x_0\|)} \leq \frac{1}{1 - u_0(\delta_0)} =: \xi.$$

Next, we shall show by induction on k that for some $\lambda \in (0,1)$

$$\beta_{k+1} \leq \lambda^{k+1}\delta \text{ for } k = 0,1,2,\ldots \tag{3.7}$$

By (A_4), $b_0 \le \delta$. Suppose that $b_k \le \delta$ and $x_k \in \overline{U}(x_0, \rho)$. We shall show by mathematical induction that $b_{k+1} \le \delta$.

Using (A_3) and (A_4), we get in turn that

$$\begin{aligned}
b_{k+1} = \|F'(x_0)^{-1}F(x_{k+1})\| &\le \|F'(x_0)^{-1}\Big(F(x_{k+1}) - F(x_k) - F'(x_k)(x_{k+1}-x_k)\Big)\| \\
&\quad + \|F'(x_0)^{-1}\Big(F(x_k) + F'(x_k)(x_{k+1}-x_k)\Big)\| \\
&\le \int_0^1 u\Big(\theta\|x_{k+1}-x_k\|\Big)d\theta\|x_{k+1}-x_k\| + (1-\alpha_k)\|F'(x_0)^{-1}F(x_k)\| \\
&\le \int_0^1 u\left(\theta\alpha_k \frac{\|F'(x_0)^{-1}F(x_k)\|}{1-W_0(\|x_k-x_0\|)}\right)d\theta\alpha_k \frac{\|F'(x_0)^{-1}F(x_k)\|}{1-W_0(\|x_k-x_0\|)} \\
&\le (1-\alpha_k)\beta_k + \alpha_k \int_0^1 u\Big(\theta\alpha_k\beta_k\xi_k\Big)d\theta\xi_k\beta_k \qquad (3.8) \\
&\le (1-\alpha_k)\delta + \alpha_k\delta = \delta,
\end{aligned}$$

which completes the induction for $\beta_{k+1} \le \delta$. Set

$$\gamma_k = \int_0^1 u\Big(\theta\alpha_k\beta_k\xi_k\Big)d\theta\beta_k\xi_k. \qquad (3.9)$$

Moreover, we obtain by (3.8) that

$$\beta_{k+1} \le \Big[1-\alpha_k+\alpha_k\int_0^1 u\Big(\theta\alpha_k\beta_k\xi_k\Big)d\theta\xi_k\Big]\beta_k \le \lambda\beta_k \le \lambda^{k+1}\beta_0 \le \lambda^{k+1}\delta, \quad (3.10)$$

which completes the induction for (3.7), where

$$0 < 1-\alpha_k+\alpha_k\int_0^1 u\Big(\theta\alpha_k\beta_k\xi_k\Big)d\theta\xi_k \le \lambda < 1. \qquad (3.11)$$

We also have that

$$\begin{aligned}
\|x_{k+1}-x_0\| &\le \|x_{k+1}-x_k\| + \cdots + \|x_1-x_0\| \\
&\le \xi\lambda^k\delta + \xi\lambda^{k-1}\delta + \cdots + \xi\lambda\delta + \|x_1-x_0\| \\
&\le \xi\lambda\delta\frac{1-\lambda^{k+1}}{1-\lambda} + \alpha_0\|F'(x_0)^{-1}F'(x_0)\| < \delta, \qquad (3.12)
\end{aligned}$$

so, $x_{k+1} \in U(x_0, \delta)$. Similarly, we have for $m = 1, 2, \dots$ that

$$\|x_{k+m} - x_k\| \leq \xi \lambda^k \beta_0 \frac{1-\lambda^m}{1-\lambda}, \tag{3.13}$$

which shows that sequence $\{x_n\}$ is complete in a Banach space X and as such it converges to some $x^* \in \overline{U}(x_0, \delta)$ (since $\overline{U}(x_0, \delta)$ is a closed set). By (A_8) and (3.9), $\lim_{k \to +\infty} \gamma_k = 0$.

Hence, by Lemma 3.1,

$$\lim_{k \to \infty} \beta_k = \lim_{k \to \infty} \|F'(x_0)^{-1} F(x_k)\| = 0 \tag{3.14}$$

so by $(A_1), F(x^*) = 0$. Therefore, $x^* \in \overline{U}(x_0, \delta)$ is a solution of equation $F(x) = 0$.

Finally, to show the uniqueness part, Let $y^* \in D_3$ with $F(y^*) = 0$. Using (A_2) and (A_6), we get that

$$\begin{aligned} \|F'(x_0)^{-1}(Q - F'(x_0))\| &\leq \int_0^1 u_0\Big(\theta\|x^* - x_0\| + (1-\theta)\|y^* - x_0\|\Big) d\theta \\ &\leq \int_0^1 u_0\Big(\theta_\delta + (1 - \theta_\delta^*)\Big) d\theta < 1, \end{aligned} \tag{3.15}$$

so Q is invertible. The rest is given in Theorem 2.1.

□

Remark 3.3. Lemma 3.1 also holds for $n \geq N_0$, where N_0 is a fixed integer greater than zero. In this case Theorem 3.2 also holds recording the sequence $\{x_n\}$ as follows $y_i = x_{N_0} + i$, for $i = 0, 1, 2, \dots$, and using sequence $\{y_i\}$ instead of $\{x_n\}$ in Theorem 3.2

4. Numerical Example

Consider the function $f(x)$ defined on $D = X = Y = \mathbb{R}$ by

$$f(x) = x^3 - 3x + 3, \; x_0 = 0$$

and choose

$$\alpha_n = \frac{1}{\sqrt[8]{\frac{5n+1}{10}}}.$$

Comparison Table 1.

n	Newton's Method (1.2) x_n	Quasi - Newtonian's Method (1.4) x_n
0	0	0
1	1	1.333521432163324
2	DNE	0.707698641704208
3	DNE	1.520234800970218
4	DNE	1.052105494697620
5	DNE	-1.812745521054344
6	DNE	-2.133838804017564
7	DNE	-2.108232568993786
8	DNE	-2.104468945970597
9	DNE	-2.103911245680715
10	DNE	-2.103822137361416
11	DNE	-2.103806854884213
12	DNE	-2.103804071551862
13	DNE	-2.103803538044450
14	DNE	-2.103803431167666
15	DNE	-2.103803408913973
16	DNE	-2.103803404119103
17	DNE	-2.103803403053883
18	DNE	-2.103803402810613
19	DNE	-2.103803402753646
20	DNE	-2.103803402739996
21	DNE	-2.103803402736656
22	DNE	-2.103803402735823
23	DNE	-2.103803402735611
24	DNE	-2.103803402735556
25	DNE	-2.103803402735542
26	DNE	-2.103803402735538
27	DNE	-2.103803402735537

The conditions that we have provided for α_n are sufficient, so the method may still converge even, if some of the requirements for α_n are not satisfied. Thus, looking at the table, we can conclude that Newton's method for equation $f(x) = x^3 - 3x + 3$, is not converging whenever $x_0 = 0$ while Newtonian Quasi method converges after some iterations.

References

[1] Argyros, I.K., *Convergence and Applications of Newton-type Iterations*, Springer Verlag, Berlin, Germany, 2008.

[2] Argyros, I.K. and Magreñán, A.A., *A Contemporary Study of Iterative Methods*, Academic Press, Elsevier, San Diego, CA, USA, 2017.

[3] Kantorovich, L.V. and Akilov, G.P., *Functional Analysis in Normed Spaces*, Pergamon Press, New York, 1964.

[4] Mann, W.R., Mean Value in Iteration, *Proc. Amer. Math. Soc 4*, (1953), 506-510.

[5] Pavaloiu, I., Rezolvarea ecuaiilor prin interpolare, *Ed. Ducia*, 1981 (In Romannian)

[6] Weng, X., Fixed Point Iteration for Local Strictly Pseudo Contractive Mapping, *Proc. Amer. Math.* Soc. 113, (1991), 727-731.

Chapter 5

Extended Ball Convergence for a Chebyshev-Halley Family of Methods with One Parameter

1. Introduction

Let X,Y be Banach spaces and $D \subset X$ be nonempty, open and convex set. We denote by $L(X,Y)$ the space of bounded linear operators from X into Y. Let also $U(u,a), \overline{U}(u,a)$ stand for the open and closed balls in X, respectively with center $u \in X$ and or radius $a > 0$.

Numerous problems in diverse disciplines such as Applied Mathematics, Mathematical Physics, Chemistry, Biology, Economics and also in Engineering can be brought in a form like

$$F(x) = 0, \tag{1.1}$$

Using Mathematical Modeling [2,5,6,16,14] where $F : D \to Y$ is continuously differentiable in the sense of Fréchet. The task of finding a locally unique solution of $p \in D$ in closed form can only be achieved in special cases. That is why most researcher and practitioners develop iterative methods generating a sequence approximating p.

In this chapter, we are concerned with the local convergence analysis of Chebysher-Halley Method (CHM) with parameter α defined for $n = 0, 1, 2, \ldots$ by

$$y_n = x_n - F'(x_n)^{-1}F(x_n)$$
$$x_{n+1} = y_n + \frac{1}{2}\Big(I - \alpha H(x_n)\Big)^{-1} H(x_n)(y_n - x_n), \tag{1.2}$$

where x_0 is an initial point,

$$H(x) = F'(x)^{-1}F''(x)F'(x)^{-1}F(x), \text{ and } \alpha \in \mathbb{R} \text{ or } \alpha \in \mathbb{C}$$

The study of convergence of iterative methods is usually centered into two categories: semi-local and local convergence analysis. The semi-local convergence is based on the information around an initial point, to obtain conditions ensuring the convergence of these algorithms, while the local convergence is based on the information around a solution to find estimates of the computed radii of the convergence balls. Local results are important since they provide the degree of difficulty in choosing initial points.

The local convergence analysis of CHM was given in [14] using average Lipschitz conditions. However, the radius of convergence is small in general. In this chapter,, we present a technique that can be used on other iterative methods where, we locate a subset D_0 of D containing the iterates leading to tighter majorizing sequences and a larger convergence radius using the same functions as in [14].

The rest of the chapter is structured as follows: Section 2 contains the local convergence analysis of CHM, where the numerical examples appear in Section 3.

2. Local Convergence

We shall estimate the radius of convergence for CHM using some functions and parameters. Let w_0 be a real continuous and nondecreasing function defined on $\mathbb{R}_+ U\{0\}$ with $W_0(0) = 0$. Suppose that equation has at least one positive solution. Denote by δ the smallest such solution.

$$W_0(t) = 1. \tag{2.1}$$

Let $\alpha \in \mathbb{R}$ or $\alpha \in \mathbb{C}$, V, V_1 and W, W_2 be real continuous and nondecreasing function defined on the interval $[0, \delta)$. Moreover, define functions h_α and $\overline{h}_\alpha$ on $[0, \delta)$ by

$$h_\alpha(t) = \frac{V_1(t) \int_0^1 V(\theta t) d\theta t}{(1 - W_0(t))^2} |\alpha|$$

and

$$\overline{h}_\alpha(t) = h_\alpha(t) - 1.$$

we have $\overline{h}_\alpha(\theta) = -1 < 0$ and $\overline{h}_\alpha(t) \to \infty$ as $t \to \delta^-$. The intermediate value theorem guarantees the existence of at least one solution of the equation $\overline{h}_\alpha(t) = 0$ on $(0, \delta)$.
Denote by δ_α^- the smallest such solution. Let $s > 1$, if $W_2(0) \neq 0$ and $s \geq 1$, if $W_2(0) = 0$.
Furthermore, define function φ_α and $\overline{\varphi}_\alpha$ on the interval $[0, \overline{\delta}_\alpha]$ by

$$\varphi_\alpha(t) = \frac{W_2(t)t^{s-1} + \frac{V_1(t)\Omega}{2\left(1 - W_0(t)\right)}\left(1 + \frac{\int_0^1 V(\theta t) d\theta}{1 - W_0(t)}\right) + |\alpha| V_1(t) \int_0^1 V(\theta t) d\theta \Omega}{\left(1 - h_\alpha(t)\right)\left(1 - W_0(t)\right)}$$

and

$$\overline{\varphi}_\alpha(t) = \varphi_\alpha(t) - 1.$$

where,

$$\Omega = \int_0^1 W\left((1-\theta)t\right) d\theta t$$

We get $\overline{\varphi}_\alpha(0) = -1$ and $\overline{\varphi}_\alpha(t) \to \infty$ as $t \to \overline{\delta}_\alpha^-$. Denote by δ_α be the smallest solution of equation $\overline{\varphi}_\alpha(t) = 0$ on $(0, \overline{\delta}_\alpha)$. Then, we have that for each $t \in [0, \delta_\alpha)$

$$0 \leq W_0(t) < 1, \tag{2.2}$$

$$0 \leq h_\alpha(t) < 1 \tag{2.3}$$

and

$$0 \leq \varphi_\alpha(t) < 1. \tag{2.4}$$

We also have that for $|\alpha_1| \leq |\alpha_2|$

$$\varphi_{\alpha_1}(t) \leq \varphi_{\alpha_2}(t) \tag{2.5}$$

and

$$\delta_{\alpha_2} \leq \delta_{\alpha_1}. \tag{2.6}$$

Hence, the radius of convergence increases with δ_0 being the largest.

Lemma 2.1. (Banach Lemma) [16] If $T \in L(X,Y)$ and $\|T\| < 1$, then $(I-T)^{-1} \in L(Y,X)$ and

$$\|(I-T)^{-1}\| \leq \frac{1}{1-\|T\|}$$

Lemma 2.2 [16] If $T_1 \in L(Y,X), T_2 \in L(X,Y)$ such that T_2^{-1} exists and $\|I - T_2T_1\| < 1$, then and T_1^{-1} exists

$$\|T_1^{-1}T_2\| \leq \frac{1}{1-\|I-T_2^{-1}T_1\|}$$

Lemma 2.3. If $T \in L(X,Y)$ such that $(I-T)^{-1}$ exists, then

$$T(I-T)^{-1} = (I-T)^{-1}T$$

Proof. We have

$$T - T^2 = T - T^2 \Rightarrow (I-T)T = T(I-T)$$

$$T = (I-T)^{-1}T(I-T)$$

$$\Rightarrow T(I-T)^{-1} = (I-T)^{-1}T(I-T)(I-T)^{-1}$$

$$\Rightarrow T(I-T)^{-1} = (I-T)^{-1}T.$$

□

We need an Ostrowski-type representation for CHM:

Lemma 2.4 Suppose: p is a solution of equation

$$F(x) = 0;$$

Iterates $\{x_n\}$ generated by CHM are well defined for all $n = 0,1,2,\ldots$. Then, the following representation is true for all $n = 0,1,2,\cdots$.

$$x_{n+1} - p = \left(I - \alpha H(x_n)\right)^{-1} F'(x_n)^{-1} \Big[F(p) - F(x_n) - F'(x_n)(p - x_n) - \frac{1}{2}F''(x_n)(p - x_n)(p - x_n) + \frac{1}{2}\lambda_n + \alpha\mu_n\Big], \tag{2.7}$$

where

$$y_n = x_n - F'(x_n)^{-1}F(x_n) \tag{2.8}$$

$$\begin{aligned}\lambda_n &= F''(x_n)(p - x_n)(p - x_n) - F''(x_n)(y_n - x_n)(y_n - x_n)\\ &= F''(x_n)(p - x_n)(p - y_n) + F''(x_n)(y - x_n)(p - y_n)\end{aligned} \tag{2.9}$$

and

$$\begin{aligned}\mu_n &= F''(x_n)(y_n - x_n)(y_n - x_n) + F''(x_n)(y_n - x_n)(x_n - p)\\ &= F''(x_n)(y_n - x_n)(y_n - p).\end{aligned} \tag{2.10}$$

Proof. We can write by CHM, and since by Lemma 2.3

$$H(x_n)\left(I - \alpha H(x_n)\right)^{-1} = (I - \alpha H(x_n))^{-1} H(x_n),$$

$$\begin{aligned}x_{n+1} - p &= x_n - p + (y_n - x_n) + \frac{1}{2}(I - \alpha H(x_n))^{-1} H(x_n)(y_n - x_n),\\
&= (I - \alpha H(x_n))^{-1}\Big[(I - \alpha H(x_n))(x_n - p) + (I - \alpha H(x_n))(y_n - x_n)\\
&\qquad + \frac{1}{2}H(x_n)(y_n - x_n)\Big]\\
&= (I - \alpha H(x_n))^{-1}\Big[(I - \alpha H(x_n))(x_n - p) + (I - \alpha H(x_n))(y_n - x_n)\\
&\qquad - \frac{1}{2}F'(x_n)^{-1}F''(x_n)(y_n - x_n)(y_n - x_n)\Big]\\
&= (I - \alpha H(x_n))^{-1}F'(x_n)^{-1}\Big[F'(x_n)\left(I - \alpha H(x_n)\right)(x_n - p)\\
&\qquad + F'(x_n)\left(I - \alpha H(x_n)(y_n - x_n) - \frac{1}{2}F''(x_n)(y_n - x_n)(y_n - x_n)\right].\end{aligned} \tag{2.11}$$

The preceding bracket can be simplified even further, using

$$F'(x_n)\left(I - \alpha H(x_n)\right) = F'(x_n) + \alpha F''(x_n)(y_n - x_n),$$

and the definition of y_n, λ_n, μ_n to obtain

$$\begin{aligned}
&F'(x_n)(x_n-p)+\alpha F''(x_n)(p-x_n)(x_n-p)+F'(x_n)(y_n-x_n)\\
&\qquad +\alpha F''(x_n)(y_n-x_n)(y_n-x_n)-\frac{1}{2}F''(x_n)(y_n-x_n)(y_n-x_n)\\
&=F'(x_n)(x_n-p)+\alpha\big[F''(x_n)(y_n-x_n)(x_n-p)+F''(x_n)(y_n-x_n)(y_n-x_n)\big]\\
&\qquad +F'(x_n)(-F'(x_n)^{-1}F(x_n)-\frac{1}{2}F''(x_n)(y_n-x_n)(y_n-x_n)\\
&=F'(x_n)(x_n-p)+\alpha\mu_n-F(x_n)-\frac{1}{2}\big[F''(x_n)(p-x_n)(p-x_n)-\lambda_n\big]\\
&=F(p)-F(x_n)-F'(x_n)(p-x_n)-\frac{1}{2}F''(x_n)(p-x_n)(p-x_n)+\frac{1}{2}\lambda_n+\alpha\mu_n.
\end{aligned} \tag{2.12}$$

Then, if we replace the bracket in (2.11) by the right hand side of (2.12), we obtain (2.7). □

Let $u \in X$ and $r > 0$. The set $U(u,r) = \{x \in X \mid \|x-u\| < r\}$ is called the open ball in X with center u and radius $r > 0$. The set $\overline{U}(u,r)$ is the closure of $\overline{U}(u,r)$.

The local convergence of CHM is based on hypotheses (A):

(A_1) $F : D \subseteq X \to Y$ is twice continuously Fréchet-differentiable.

(A_2) There exists $p \in D$ such that $F(p) = 0$ and $F'(p)^{-1} \in L(X,Y)$.

(A_3) There exists a real continuous and nondecreasing function W_0 defined on $\mathbb{R}+U\{0\}$ with $W_0(0) = 0$ such that for each $x \in D$

$$\|F'(p)^{-1}(F'(x)-F'(p))\| \leq W_0(\|x-p\|).$$

Set $D_0 = D \cap U(p,\delta)$, where δ is given in (2.1).

(A_4) There exist continuous and nondecreasing function W, W_2, V, V_1 defined on the interval $[0,\delta)$ and $s > 1$, if $W_2(0) = 0$ or $s \geq 1$, if $W_2(0) = 0$ such that for each $x, y \in D_0$

$$\|F'(p)^{-1}(F'(x)-F'(y))\| \leq W(\|x-p\|),$$

$$\|F'(p)^{-1}\left[F(p)-F(x)-F'(x)(p-x)-\tfrac{1}{2}F''(x)(p-x)(p-x)\right]\|$$

$$\leq W_2(\|x-p\|)\|x-p\|^s,$$

$$\|F'(p)^{-1}F'(x)\| \leq V(\|x-p\|)$$

and

$$\|F'(p)^{-1}F''(x)\| \leq V_1(\|x-p\|).$$

(A_5) $\overline{U}(p,\delta_\alpha) \subseteq D$, where δ_α is defined previously.

Next, we are in a position to show the local convergence analysis using previously introduced auxiliary results hypotheses (A), and notation.

Theorem 2.5 Suppose that the hypotheses (A) hold. Then, iteration $\{x_n\}$ starting from $x_0 \in U(p,\delta_\alpha)-\{p\}$ and produced by CHM is well defined in $U(p,\delta_\alpha)$, stays in $U(p,\delta_\alpha)$ for all $n=0,1,2,\ldots$, and converges to p so that

$$\|x_{n+1}-p\| \leq \varphi_\alpha(\|x_n-p\|)\|x_n-p\| \leq \|x_n-p\| < \delta_\alpha, \tag{2.13}$$

where the function φ_α is defined previously. Moreover, if there exists $r \geq \delta_\alpha$ such that

$$\int_0^1 W_0(\theta r)\, d\theta < 1, \tag{2.14}$$

then the limit point p is the only solution of equation $F(x)=0$ in $D_1 = D \cap \overline{U}(p,r)$.

Proof. We base the proof on induction. By hypothesis $x_0 \in U(p,\delta_\alpha)-\{p\},(A_2),(A_3)$ and (2.1), we get in turn

$$\|F'(p)^{-1}(F'(x_0)-F'(p))\| \leq W_0(\|x_0-p\|) \leq W_0(\delta_\alpha) < 1. \tag{2.15}$$

It then follows from Lemma 2.2 and (2.15) that $F'(x_0)^{-1} \in L(Y,X)$ and

$$\|F'(x_0)^{-1}F'(p)\| \leq \frac{1}{1-W_0(\|x_0-p\|)}. \tag{2.16}$$

We can write by (A_2) that

$$F(x_0) = F(x_0) - F(p) = \int_0^1 F'(p)^{-1}F'\Big(p+\theta(x_0-p)\Big)(x_0-p)\,d\theta \qquad (2.17)$$

Notice that $\|p+\theta(x_0-p)-p\| = \theta\|x_0-p\| \leq \delta_\alpha$ for all $\theta \in [0,1]$, so $p+\theta(x_0-p) \in U(p,\delta_\alpha)$. By (A_4) and (2.17), we get that

$$\begin{aligned}\|F'(p)^{-1}F(x_0)\| &= \|\int_0^1 F'(p)^{-1}F'\Big(p+\theta(x_0-p)\Big)(x_0-p)d\theta\| \\ &\leq \int_0^1 V(\|x_0-p\|)\|x_0-p\|\,d\theta. \qquad (2.18)\end{aligned}$$

In view of (A_4), (2.16) and (2.18), we obtain that

$$\begin{aligned}\|\alpha H(x_0)\| &= \|\alpha F'(x_0)^{-1}F''(x_0)F'(x_0)^{-1}F(x_0)\| \\ &\leq |\alpha|\,\|F'(x_0)^{-1}F'(p)\|\,\|F'(p)^{-1}F''(x_0)\|\,\|F'(x_0)^{-1}F'(p)\|\,\|F'(p)^{-1}F(x_0)\| \\ &\leq |\alpha|\frac{V_1(\|x_0-p\|)\int_0^1 V(\theta\|x_0-p\|)\|x_0-p\|\,d\theta}{\Big(1-W_0(\|x_0-p\|)\Big)^2} \\ &= h_\alpha(\|x_0-p\|) \leq h_\alpha(\delta_\alpha) < 1, \qquad (2.19)\end{aligned}$$

so $(I-\alpha H(x_0))^{-1} \in L(X,Y)$,

$$\|(I-\alpha H(x_0))^{-1}\| \leq \frac{1}{1-h_\alpha(\|x_0-p\|)}, \qquad (2.20)$$

and x_1 is well defined by CHM for $n = 0$. Next, we shall make use of Lemma 2.3 and Lemma 2.4 to show (2.13) for $n = 0$.

Indeed, using (2.4), (2.7) - (2.10), (2.16), (2.18) and (2.20), we have in turn that

$$\begin{aligned}
\|x_1-p\| &\leq \|(I-\alpha H(x_0))^{-1}\|\,\|F'(x_0)^{-1}F'(p)\|\Big[\|F'(p)^{-1}\Big(F(p)-F(x_0)-F'(x_0)(p-\\
&\qquad x_0)-\frac{1}{2}F''(x_0)(p-x_0)(p-x_0)\Big)\|+\frac{1}{2}\|F'(p)^{-1}\lambda_0\|+|\alpha|\,\|F'(p)^{-1}\mu_0\|\Big]\\
&\leq \|(I-\alpha H(x_0))^{-1}\|\,\|F'(x_0)^{-1}F'(p)\|\Big[\|F'(p)^{-1}\Big(F(p)-F(x_0)-F'(x_0)(p-x_0)\\
&\qquad -\frac{1}{2}F''(x_0)(p-x_0)(p-x_0)\Big)\|+\frac{1}{2}\|F'(p)^{-1}F''(x_0)(\|p-x_0\|\,\|p-y_0\|\\
&\qquad +\|y_0-x_0\|\,\|p-y_0\|)+|\alpha|\,\|F'(p)^{-1}F''(x_0)\|\,\|y_0-x_0\|\,\|y_0-p\|\Big]\\
&\leq \frac{W_2(\|x_0-p\|)\|x_0-p\|^s+A+R}{\Big(1-h_\alpha(\|x_0-p\|)\Big)\Big(1-W_0(\|x_0-p\|)\Big)}\\
&= \varphi_\alpha(\|x_0-p\|)\|x_0-p\| \leq \|x_0-p\| \leq \delta_\alpha,
\end{aligned} \tag{2.21}$$

where,

$$A=\frac{\frac{1}{2}V_1(\|x_0-p\|)\int_0^1 W\Big((1-\theta)\|x_0-p\|\Big)d\theta\|x_0-p\|^2}{1-W_0(\|x_0-p\|)}\left(1+\frac{\int_0^1 V(\theta\|x_0-p\|)d\theta}{1-W_0(\|x_0-p\|)}\right)$$

and,

$$R=|\alpha|\;V_1(\|x_0-p\|)\frac{\int_0^1 V(\theta\|x_0-p\|)d\theta\int_0^1 W\Big((1-\theta)\|x_0-p\|\Big)d\theta\|x_0-p\|^2}{\Big(1-W_0(\|x_0-p\|)\Big)^2},$$

which shows (2.13) for $n=0$ and $x_1\in U(p,\delta_\alpha)$.

The induction for (2.3) is completed, if we replace x_0,x_1 by x_k,x_{k+1} in the preceding estimates. Then, in view of the estimate

$$\|x_{k+1}-p\| \leq c\|x_k-p\| < \delta_\alpha, \tag{2.22}$$

$c=\varphi_\alpha(\|x_0-p\|)\in[0,1)$, we deduce that

$$\lim_{k\to+\infty} x_k = p \text{ and } x_{k+1}\in U(p,\delta_\alpha)$$

The uniqueness of p is shown by letting $p^* \in D_1$ with $F(p^*) = 0$ and setting

$$Q = \int_0^1 F'(p^* + \theta(p - p^*))d\theta.$$

It then follows from (A_3) and (2.14) that

$$\|F'(p)^{-1}(Q - F'(p^*))\| \leq \int_0^1 W_0(\theta\|p - p^*\|)d\theta \leq \int_0^1 W_0(\theta r)d\theta < 1, \quad (2.23)$$

so, $Q^{-1} \in L(Y, X)$. Using the identity

$$0 = F(p^*) - F(p) = Q(p^* - p),$$

we conclude that $p^* - p$.

□

Remark 2.6 (a) Let us consider the popular choices for functions $W_0, W, W_2, V,$ and V_1 given by

$$W_0(t) = L_0 t, W(t) = Lt, W_2(t) = \frac{M}{6}t^2, s = 1, V(t) = K \text{ and } V_1(t) = L,$$

corresponding to the Lipschitz-type hypotheses,

$$\|F'(p)^{-1}(F'(x) - F'(p))\| \leq L_0\|x - p\| \text{ for each } x \in D, \quad (2.24)$$

$$\|F'(p)^{-1}(F'(x) - F'(y))\| \leq L\|x - p\| \text{ for each } x, y \in D_0, \quad (2.25)$$

$$\|F'(p)^{-1}(F(x) - F(y))\| \leq K\|x - y\| \text{ for each } x, y \in D_0, \quad (2.26)$$

and

$$\|F'(p)^{-1}(F''(x) - F''(p))\| \leq M\|x - p\| \text{ for each } x, y \in D_0. \quad (2.27)$$

Then, it can easily be seen by (2.13) that there exists $c_0 > 0$ such that

$$\|x_{n+1} - p\| \leq c_0\|x_n - p\|^3, \quad (2.28)$$

from which, we obtain the order of convergence three for HCM.
Notice that in the literature hypotheses for each $x, y \in D$

$$\|F'(p)^{-1}(F'(x) - F'(y))\| \leq \overline{L}\|x - p\| \tag{2.29}$$

$$\|F'(p)^{-1}(F'(x) - F'(y))\| \leq \overline{K}\|x - y\| \tag{2.30}$$

and

$$\|F'(p)^{-1}(F''(x) - F''(y))\| \leq \overline{M}\|x - p\| \tag{2.31}$$

are used instead of (2.24)-(2.27). However, we have

$$L_0 \leq L \leq \overline{L}, \tag{2.32}$$

$$K \leq \overline{K} \tag{2.33}$$

and

$$M \leq \overline{M} \tag{2.34}$$

hold in general, since $D_0 \subseteq D$. Hence, the new radius of convergence δ_α is at least large as the ones using $L = \overline{L} = L_0, \overline{K}$ and $\overline{M}$ denoted by δ_α^0, i.e.

$$\delta_\alpha^0 \leq \delta_\alpha. \tag{2.35}$$

The error bounds are also at least as precise.

(b) The results in Schewetlick's book, Zhendga et all [14] and other related works [1-13,15-17] can be immediately improved if Do replaces D in those studies (See also the numerical examples).

Remark 2.7 The results of Theorem 2.5 can be extended to the following class parameterized Chebysher-Halley method (PCHM) parameterized by scalar α and by iteration i defined in Schwetlick's book by

$$y_{n,i+1} = x_n - A_{n,i}^{-1}\left[\frac{1-\alpha}{2}F''(x_n)(y_{n,i} - x_n)(y_{n,i} - x_n) + F(x_n)\right], \tag{2.36}$$

where

$$y_{n,0} = x_n, y_{n,i+1} = x_{n+1}$$

and

$$A_{n,i} = F'(x_n) + \frac{\alpha}{2}F''(x_n)(y_{n,1} - x_n).$$

It was shown in Schwetlick's book that, if $i = 1$ in (2.36), then (2.36) reduces to (1.2).

Next, we shall present the local convergence of (PCHM) along the lines of CHM and Theorem 2.5. But first, we need some scalar functions and parameters. Define functions Ψ_0 and $\overline{\Psi}_0$ on $[0, \delta_\alpha)$ by

$$\Psi_0(t) = W_0(t) + |\alpha| V_1(t) t$$

and

$$\overline{\Psi}_0(t) = \Psi_0(t) - 1.$$

We have that $\Psi_0(0) = -1$.
Suppose that

$$\overline{\Psi}_0(t) \longrightarrow +\infty \text{ or a positive number as } t \longrightarrow \delta_\alpha^-. \tag{2.37}$$

Denote by r_α^0 the smallest solution of equation $\overline{\Psi}_0(t) = 0$ in $(0, \delta_\alpha)$. Moreover, define functions Ψ and $\overline{\Psi}$ on $[0, r_\alpha^0)$ by

$$\Psi(t) = \frac{1}{1 - \Psi_0(t)} \Big[\int_0^1 W((1-\theta)t) d\theta + |\alpha| V_1(t) t + 2|1-\alpha| V_1(t) t \Big]$$

and

$$\overline{\Psi}(t) = \Psi(t) - 1.$$

We get that $\overline{\Psi}(0) = -1$ and $\Psi(t) \longrightarrow +\infty$ as $t \longrightarrow (r_\alpha^0)^-$. Denote by r_α the smallest solution of equation $\overline{\Psi}(t) = 0$ in $(0, r_\alpha^0)$.

In view of (2.13), (2,36), the definition of δ_α and r_α, we have that

$$\|x_{n+1} - p\| = \|y_{n,2} - p\| \le \varphi_\alpha(\|y_{n,0} - p\|) \|y_{n,0} - p\| \le \|y_{n,0} - p\| \le \|x_n - o\| < r. \tag{2.38}$$

We also get by (2.38), the definition of function Ψ_0 and radius r_α that

$$\begin{aligned}\|F'(p)^{-1}(A_{n,2}-F'(p))\| &\leq \|F'(p)^{-1}(F'(x_n)-F'(p))\| + \frac{|\alpha|}{2}\|F'(p)^{-1}F''(x_n)(y_{n,2}-x_n)\| \\ &\leq W_0(\|x_n-p\|) + \frac{|\alpha|}{2}V_1(\|x_n-p\|)(\|y_{n,2}-p\|+\|x_n-p\|) \\ &\leq W_0(\|x_n-p\|) + \frac{|\alpha|}{2}V_1(\|x_n-p\|)(1+\varphi_\alpha(\|y_{n,0}-p\|)\|x_n-p\|) \\ &\leq W_0(\|x_n-p\|) + |\alpha|V_1(\|x_n-p\|)\|x_n-p\| \\ &= \Psi_0(\|x_n-p\|) \leq \Psi_0(r_\alpha) < 1, \end{aligned} \tag{2.39}$$

so $A_{n,2}^{-1}$ exists and

$$\|A_{n,2}^{-1}F'(p)\| \leq \frac{1}{1-\Psi_0(\|x_n-p\|)}. \tag{2.40}$$

Moreover, $y_{n,3}$ exists by (2.36) for $i=2$. We can write by (2.36) for $i=2$

$$\begin{aligned} y_{n,3}-p &= x_n - p - A_{n,2}^{-1}\Big[\frac{1-\alpha}{2}F''(x_n)(y_{n,2}-x_n)(y_{n,2}-x_n)+F(x_n)\Big] \\ &= A_{n,2}^{-1}\Big[(F'(x_n)(x_n-p)-(F(x_n)-F(p))) + \frac{\alpha}{2}F''(x_n)(y_{n,2}-x_n)(x_n-p) \\ &\quad - \frac{1-\alpha}{2}F''(x_n)(y_{n,2}-x_n)(y_{n,2}-x_n)\Big]. \end{aligned} \tag{2.41}$$

Using (A_4), (2.40) and (2.41), we obtain in turn that

$$\begin{aligned} \|y_{n,3}-p\| &\leq \|A_{n,2}^{-1}F'(p)\|\Big[\|F'(p)^{-1}(F'(x_n)(x_n-p)-(F(x_n)-F(p)))\| \\ &\quad + \frac{|\alpha|}{2}\|F'(p)^{-1}F''(x_n)\|\;\|(y_{n,2}-p)+(p-x_n)\|\;\|x_n-p\| \\ &\quad + \frac{|1-\alpha|}{2}\|F'(p)^{-1}F''(x_n)\|\|(y_{n,2}-p)+(p-x_n)\|^2\Big] \\ &\leq \frac{1}{1-\Psi_0(\|x_n-p\|)}\Big[\int_0^1 W((1-\theta)\|x_n-p\|)d\theta + |\alpha|V_1(\|x_n-p\|)\|x_n-p\| \\ &\quad + 2|1-\alpha|V_1(\|x_n-p\|)\|x_n-p\|\Big]\|x_n-p\| \\ &= \Psi(\|x_n-p\|)\|x_n-p\| \leq \|x_n-p\| < r_\alpha, \end{aligned} \tag{2.42}$$

which shows $y_{n,3} \in U(p, r_\alpha)$ and

$$\|y_{n,i+1}-p\| \leq \Psi(\|x_n-p\|)\|x_n-p\| \leq \|x_n-p\| < r_\alpha \tag{2.43}$$

holds for $i = 2$. By replacing $y_{n,2}, y_{n,3}$ by $y_{n,k}, y_{n,k+1}$ in the preceding estimates we show (2.43) holds for $i \geq 2$. Then, from the estimate

$$\|x_{n+1} - p\| = \|y_{n,i+1} - p\| \leq \overline{c}\|x_n - p\| < r_\alpha, \overline{c} = \Psi(\|x_0 - p\|) \in [0,1), \quad (2.44)$$

we conclude that

$$\lim_{n\to+\infty} y_{n,i+1} = p$$

holds for all $i = 1, 2, \ldots$.
Hence, we arrived at:

Proposition 2.8 Suppose the hypotheses of Theorem 2.5 and (2.37) hold. Then, sequence $\{y_{n,i}\}$ generated for $x_0 \in U(p,r) - \{p\}, i = 1, 2, \ldots$ by PCHM is well defined, remains in $U(p, r_\alpha)$ and converges to p so that estimate (2.43) is satisfied, where r_α is defined previously. Moreover, if there exists $\overline{r} \geq r_\alpha$ such that

$$\int_0^1 W_0(\theta\overline{r})d\theta < 1,$$

Then, the limit point p is the only solution of equation $F(x) = 0$ in $D_2 = D \cap \overline{U}(p, \overline{r})$.

3. Numerical Example

Example 3.1 Suppose that the motion of an object in three dimensions is governed by the system of differential equation

$$\begin{aligned} f_1'(x) - f_1(x) &= 0 \\ f_2'(y) - (e-1)y - 1 &= 0 \\ f_3'(z) - 1 &= 0, \end{aligned}$$

with $x, y, z \in D$ for $f_1(0) = f_2(0) = f_3(0) = 0$. Then, the solution of the system is given by $W = (x, y, z)^T$ by the function $F : (f_1, f_2, f_3) : D \to \mathbb{R}^3$ defined by

$$F(V) = \left(e^x - 1, \frac{e-1}{2}y^2 + y, z\right)^T.$$

The Fréchet derivative is given by

$$F'(V) = \begin{bmatrix} e^x & 0 & 0 \\ 0 & (e-1)y+1 & 0 \\ 0 & 0 & 1 \end{bmatrix}$$

Then, using hypothesis followed by (A_4) and (A_3), we get,

$$\begin{aligned} W_0(t) &= (e-1)t, \\ W(t) &= e^{\frac{1}{e-1}}t, \\ W_2(t) &= \frac{e^{\frac{1}{e-1}}}{2}, \\ V_1(t) &= V(t) = e^{\frac{1}{e-1}}. \end{aligned}$$

Choosing $\alpha = 0.1, \delta = 2$, and using equations followed by (2.1), we found that

$$\delta_{0.1}^{-1} \approx 0.379173851596588,$$

and

$$\delta_{0.1} \approx 0.227932841756967.$$

Moreover, choosing $\alpha = 0.2, \delta = 2$, and using equations followed by (2.1), we found that

$$\delta_{0.2}^{-1} \approx 0.318936565386782,$$

and

$$\delta_{0.2} \approx 0.210758240039844.$$

which further satisfies (2.5) and (2.6), where $0.1 < 0.2$ followed by $\delta_{0.2} < \delta_{0.1}$.

References

[1] Amat, S., Busquier, S. and Gutierrez, J.M., Geometric constructions of iterative functions to solve nonlinear equations, *J. Comput. Appl. Math.* 157, 2003, 197-205.

[2] Argyros, I.K., *Convergence and Applications of Newton-type Iterations*, Springer Verlag, Berlin, Germany, 2008.

[3] Argyros, I.K. and Chen, D., Results on the Chebyshev method in Banach spaces. *Proyecciones* 12, 1993, 119-128.

[4] Argyros, I.K. and Magreñán, A.A., *A Contemporary Study of Iterative Methods*, Academic Press, Elsevier, San Diego, CA, USA, 2017.

[5] Argyros, I.K., Chen, D. and Qian, Q., A convergence analysis for rational methods with a parameter in Banach space, *Pure Math. Appl.* 5, 1994, 59-73.

[6] Argyros, I.K., On an improved unified convergence analysis for a certain class of Euler-Halley type methods, *J. Korean Soc. Math. Educ. Ser. B Pure Appl. Math.* 13, 2006, 207-215.

[7] Chen,D., Argyros, I.K. and Qian, Q., A local convergence theorem for the Supper-Halley method in a Banach space, *Appl. Math. Lett.* 7, 1994, 49-52.

[8] Ezquerro, J.A. and Hernandez, M.A., Halley's method for operators with unbounded second derivative, *Appl. Numer. Math.* 57, 2007, 354-360.

[9] Ezquerro, J.A. and Hernandez, M.A., New Kantorovich-type conditions for Halley's method, *Appl. Number. Anal. Comput. Math.* 2, 2005, 70-77.

[10] Ezquerro, J.A. and Hernandez, M.A., On the R-order of the Halley method, *J. Math. Anal. Appl.* 303, 2005, 591-601.

[11] Gutierrez, J.M. and Hernandez, M.A., A family of Chebyshev-Halley type methods in Banach space, *Bull. Aust. Math. Soc.* 55, 1997, 113-130.

[12] Gutierrez, J.M. and Hernandez, M.A., An acceleration of Newton's method: Super-Halley method, *Appl. Math. Comput.* 117, 2001, 223-239.

[13] Hernandez, M.A. and Salanova, M.A., A family of Chebyshev-Halley type methods, *Int. J. Comput. Math.* 47, 1993, 59-63.

[14] Huang, Z. and Guochun, M., On the local convergence of a family of Euler-Halley iterations with a parameter, *Numer. Algor.* 52, 2009, 419-433.

[15] Huang, Z., On a family of Chebyshev-Halley type methods in Banach space under weaker Smale condition, *Number. Math. JCU* 9, 2000, 37-44.

[16] Kantorovich, L.V. and Akilov, G.P. *Functional Analysis in Normed Spaces*, Pergamon Press, New York, 1964.

[17] Wang, X., Convergence of the iteration of Halley's family and Smale operator class in Banach space, *Sci. China Ser.* A 41, 1998, 700-709.

Chapter 6

Extending the Applicability of Newton's Method under the Second Fréchet-Derivative: Case I

1. Introduction

The applicability of Newton's method given in [9,13,14,15] involving the second Fréchet-derivative is extended under less computational effort.

2. Semi-Local Convergence Analysis

We first state the semi-local convergence analysis of Newton's method given in [9,13,14,15] in order to then compare with our new analysis that follows it. In particular, the comparison is given in Remark 2.3.

The semi-local convergence of Newton's method given in [9,12] was presented under the hypotheses (C):

(C_1) $F : D \subset X \to Y$ is a twice continuously Fréchet-differentiable operator.

(C_2) There exist $x_0 \in D, \delta \geq 0$ and $\eta \geq 0$ such that

$$F'(x_0) \in L(Y,X), \|F'(x_0)^{-1}F(x_0)\| \leq \eta$$

and

$$\|F'(x_0)^{-1}F''(x_0)\| \leq \delta.$$

(C_3) There exists a function $\overline{W} : [0,\infty) \to [0,\infty)$ continuous and nondecreasing such that for each $x \in D$

$$\|F'(x_0)^{-1}\Big(F''(x) - F''(x_0)\Big)\| \leq \overline{W}(\|x - x_0\|)$$

and $\overline{W}(0) = 0$.
Let $t_0 \geq 0$. Define function $\overline{g} : [t_0,\infty) \to \mathbb{R}$ by

$$\overline{g}(t) = \int_{t_0}^{t}\int_{t_0}^{\theta} \overline{W}(\tau - t_0)d\tau d\theta + \frac{\delta}{2}(t - t_0)^2 - (t - t_0) + \eta.$$

(C_4) $\overline{g}(\alpha) \leq 0$ and α is the unique positive solution of the equation $\overline{g}\,'(t) = 0$ such that $\alpha > t_0$ and $U(x_0, t^* - t_0) \subseteq D$, where $\bar{t}\,^*$ is the smallest positive solution of equation $\overline{g}(t) = 0$ in $[t_0,\infty)$.

Theorem 2.1 [4] Suppose that the hypotheses (C) are satisfied. Then, sequence $\{x_n\}$ generated for $x_0 \in D$ by Newton's method is well defined in $\overline{U}(x_0, \bar{t}\,^* - t_0)$, remains in $\overline{U}(x_0, \bar{t}\,^* - t_0)$ for each $n = 0,1,2,\ldots$ and converges to a solution $x^* \in \overline{U}(x_0, \bar{t}\,^* - t_0)$ of equation $F(x) = 0$. Moreover, the following estimates hold for each $n = 0,1,2,\ldots$

$$\|x_{n+1} - x_n\| \leq u_{n+1} - u_n$$

and

$$\|x_n - x^*\| \leq \bar{t}\,^* - u_n,$$

where

$$u_n = u_{n-1} - \frac{\overline{g}(u_{n-1})}{\overline{g}\,'(u_{n-1})}.$$

Furthermore, the solution x^* is unique in $U(x_0, \bar{t}^{**} - t_0) \cap D$, if $\bar{t}^* = \bar{t}^{**}$, provided that function $\bar{g}$ has two real zeros $\bar{t}^*$ and $\bar{t}^{**}$ such that $t_0 < \bar{t}^* \leq \bar{t}^{**}$.

Our semi-local convergence of Newton's method is based on the hypothesis (H):

(H_1) $F : D \subset X \to Y$ is a twice continuously Fréchet-differentiable operator.

(H_2) There exist $x_0 \in D, \delta \geq 0$ and $\eta \geq 0$ such that

$$F'(x_0)^{-1} \in L(Y,X), \|F'(x_0)^{-1}F(x_0)\| \leq \eta$$

and

$$\|F'(x_0)^{-1}F''(x_0)\| \leq \delta.$$

(H_3) There exists a function $V : [0,\infty) \to [0,\infty)$ continuous and nondecreasing such that for each $x \in D$

$$\|F'(x_0)^{-1}\Big(F'(x) - F'(x_0)\Big)\| \leq V(\|x - x_0\|)$$

and

$$V(0) = 0.$$

Let $t_0 \geq 0$. Assume equation $V(t) = 1$ has a minimal positive solution r_0. Let D_0 be

$$D_0 = D \cap U(x_0, r_0).$$

(H_4) There exists a function $W : [t_0, r_0) \to [0,\infty)$ continuous and nondecreasing such that for each $x \in D_0$

$$\|F'(x_0)^{-1}\Big(F''(x) - F''(x_0)\Big)\| \leq W(\|x - x_0\|)$$

and

$$W(0) = 0.$$

Let $t_0 \geq 0$. Define function $g : [t_0, \infty) \to \mathbb{R}$ by

$$g(t) = \int_{t_0}^{t}\int_{t_0}^{\theta} W(\tau - t_0)d\tau d\theta + \frac{\delta}{2}(t - t_0)^2 - (t - t_0) + \eta.$$

(H_5) $g(\gamma) \leq 0$ and γ is the unique positive solution of the equation $g'(t) = 0$ such that $\gamma \geq t_0$ and $U(x_0, t^* - t_0) \subset D$, where t^* is the smallest positive solution of equation $g(t) = 0$ in $[t_0, +\infty)$.

(H_6) $V(t) \leq g'(t) + 1$ for each $t \in [t_0, \gamma]$.

(H_7) There exists $\gamma_1 \geq \gamma$ such that

$$\int_0^1 V\big((1-\theta)\gamma + \theta\gamma_1\big)d\theta < 1.$$

set $D_1 = D \cap \overline{U}(x_0, \gamma_1)$.

Next, we present the semi-local convergence analysis of Newton's method using the preceding notation and hypotheses (H).

Theorem 2.2 Suppose that the hypothesis (H) are satisfied. Then, sequence $\{x_n\}$ generated for $x_0 \in D$ by Newton's method is well defined in $\overline{U}(x_0, t^* - t_0)$, remains in $\overline{U}(x_0, t^* - t_0)$ for each $n = 0, 1, 2, \ldots$, and converges to a solution $x^* \in \overline{U}(x_0, t^* - t_0)$ of equation $F(x) = 0$. Moreover, the following estimates holds for each $n = 0, 1, 2, \ldots$

$$\|x_{n+1} - x_n\| \leq t_{n+1} - t_n$$

and

$$\|x_n - x^*\| \leq t^* - t_n,$$

where

$$t_n = t_{n-1} - \frac{g(t_{n-1})}{g'(t_{n-1})}.$$

Furthermore, the point x^* is the only solution of equation $F(x) = 0$ in D_1.

Proof. Let $x \in U(x_0, r_0)$. Using (H_3), we have that

$$\|F'(x_0)^{-1}\Big(F'(x) - F'(x_0)\Big)\| \leq V(\|x - x_0\|) \leq V(r_0) < 1. \tag{2.1}$$

Then, it follows from (2.1) and the Banach Lemma on invertible operators [17] that $F'(x)^{-1} \in L(Y,X)$ and

$$\|F'(x)^{-1}F'(x_0)\| \leq \frac{1}{1-V(\|x-x_0\|)}. \tag{2.2}$$

By (H_6) and (2.2) we get that

$$\|F'(x)^{-1}F'(x_0)\| \leq \frac{1}{1-V(\|x-x_0\|)} \leq -\frac{1}{g'(\|x-x_0\|)}. \tag{2.3}$$

The rest of the proof until the uniqueness part follows exactly as in the proof given by Gutierrez (see [9]) and by simply noticing that the iterates $\{x_n\}$ remain in D_0 which is an at least as precise domain as D, since $D_0 \subseteq D$. To show the uniqueness part, let $y^* \in D_1$ with $F(y^*) = 0$. Define operator M by

$$M = \int_0^1 F'(x^* + \theta(y^* - x^*))d\theta.$$

Using (H_3) and (H_7) we obtain in turn that

$$\begin{aligned}
\|F'(x_0)^{-1}\Big(M - F'(x_0)\Big)\| &\leq \int_0^1 v\Big(\|x^* + \theta(y^* - x^*) - x_0\|\Big)d\theta \\
&\leq \int_0^1 v\Big((1-\theta)\|x^* - x_0\| + \theta\|y^* - x_0\|\Big) \\
&\leq \int_0^1 v\Big((1-\theta)\gamma + \theta\gamma_1\Big)d\theta < 1,
\end{aligned}$$

so $M^{-1} \in L(Y,X)$. Finally, using the identity

$$0 = F(y^*) - F(x^*) = M(y^* - x^*),$$

we deduce that $x^* = y^*$. □

Remark 2.3 **(i)** Suppose that the hypotheses of Theorem 2.1 and Theorem 2.3 hold. Then, we have for each $t \in [t_0, \alpha)$

$$W(t) \leq \overline{W}(t), \tag{2.4}$$

$$g(t) \leq \overline{g}(t), \tag{2.5}$$

$$g'(t) \leq \overline{g}\,'(t). \tag{2.6}$$

If for each $t_0 < s \leq t$

$$-\frac{g(s)}{g'(s)} \leq -\frac{\overline{g}(t)}{\overline{g}'(t)}$$

and

$$s - \frac{g(s)}{g'(s)} \leq t - \frac{\overline{g}(t)}{\overline{g}'(t)}$$

then a simple inductive argument shows that

$$t_n \leq u_n$$

$$t_{n+1} - t_n \leq u_{n+1} - u_n$$

so,

$$t^* \leq \overline{t}^*.$$

Hence, the error bounds on the distances are at least as tight, and the information on the location of the solution x^* is at least as precise.

(ii) Clearly, in view of (2.4) - (2.6), the hypotheses of Theorem 2.1 imply the hypotheses of Theorem 2.2 but not necessarily the vice versa.
For example from (2.6)

$$g(\alpha) \leq \overline{g}(\alpha) \leq 0,$$

the existence of the γ is implied.

(iii) The results obtained here can be improved even further, if we replace the set D_0 by the set $\overline{D}_0$ defined by

$$\overline{D}_0 = D \cap U(x_1, r_0 - \eta),$$

provided that $\eta < r_0$.

Notice that the iterates $\{x_n\}$ lie in $\overline{D}_0$ which is a more precise location than D_0, since $\overline{D_0} \subseteq D_0$. The new function $\overline{W}$ replacing W in hypothesis (H) and Theorem 2.2 will be at least as tight as W leading to at least as weak sufficient semi-local convergence conditions for Newton's method. It is worth noticing that the advantages (i) - (iii) are obtained under the same computational cost as in Theorem 2.1, since the computation of the Lipschitz-type condition (C_3) passes through the computation of condition (H_3), and condition (H_4) as special cases.

3. Local Convergence Analysis

We first base the local convergence analysis (not given in [9]) on the hypotheses (A):

(A_1) $F: D \subset X \to Y$ is a twice continuously Fréchet-differentiable operator.

(A_2) There exist $x^* \in D, \delta \geq 0$ such that $F(x^*) = 0, F'(x^*)^{-1} \in L(Y,X)$ and $\|F'(x^*)^{-1}F''(x^*)\| \leq \delta$.

(A_3) There exists a function $\overline{W} : [0,\infty) \to [0,\infty)$ continuous and nondecreasing such that for each $x \in D$.

$$\|F'(x^*)^{-1}\Big(F''(x) - F''(x^*)\Big)\| \leq \overline{W}(\|x - x^*\|)$$

and $\overline{W}(0) = 0$.

(A_4) $\overline{U}(x^*, r) \subseteq D$,
where the radius of convergence r is the smallest positive zero of function $\Psi(t) = 0$ in the interval $(0, r_p)$, where

$$r_p = \sup\{t \geq 0 : (\delta + \int_0^1 \overline{W}(\theta t)d\theta)t < 1\} \text{ for } \delta \neq 0 \text{ or } W(0) \neq 0,$$

$$\Psi(t) = \varphi(t) - 1,$$

$$\varphi(t) = \frac{\int_0^1 \overline{W}((1-\theta)t)(1-\theta)d\theta + \frac{1}{2}\delta}{1 - p(t)} t,$$

and

$$p(t) = \left(\delta + \int_0^1 \overline{W}(\theta t)d\theta\right)t.$$

Notice that $\Psi(0) = -1 < 0$, and $\Psi(t) \to +\infty$ as $t \to r_p^-$. It follows by the intermediate value theorem that function Ψ has zeros in the interval $(0, r_p)$. We denote by r the smallest such zero.

Theorem 3.1 Suppose that the hypotheses (A) are satisfied. Then ,sequence $\{x_n\}$ generated for $x_0 \in U(x^*, r) - \{x^*\}$ by Newton's method is well defined in $U(x^*, r)$, remains in $U(x^*, r)$ and converges to x^* which is the only solution of equation $F(x) = 0$ in $\overline{U}(x^*, r)$. Moreover, the following estimates hold

$$\begin{aligned} \|x_{n+1} - x^*\| &\leq \frac{\int_0^1 \overline{W}((1-\theta)\|x_n - x^*\|)(1-\theta)d\theta + \frac{1}{2}\delta}{1 - p(\|x_n - x^*\|)}\|x_n - x^*\|^2 \\ &\leq \|x_n - x^*\| < r. \end{aligned} \tag{3.1}$$

Proof. Let $x \in U(x^*, r)$. Using (A_2) and (A_3), we get in turn that

$$\begin{aligned} \|F'(x^*)^{-1}\left(F'(x) - F'(x^*)\right)\| &= \|F'(x^*)^{-1}\int_{x^*}^{x} F''(z)dz\| \\ &\leq \|F'(x^*)F''(x^*)\|\|x^* - x\| \\ &+ \|\int_0^1 F'(x^*)^{-1}\left(F''(x^* + \theta(x - x^*)) - F''(x^*)\right)(x - x^*)d\theta\| \\ &\leq \left(\delta + \int_0^1 \overline{W}(\theta\|x - x^*\|)d\theta\right)\|x - x^*\| \\ &= p(\|x - x^*\|) \leq p(r) < 1. \end{aligned} \tag{3.2}$$

It follows from (3.1) and the Banach Lemma on invertible operators that $F'(x)^{-1} \in L(Y, X)$ and

$$\|F'(x)^{-1}F'(x^*)\| \leq \frac{1}{1 - p(\|x - x^*\|)}. \tag{3.3}$$

In particular for $x = x_0$, point x_1 is well defined by Newton's method for $n = 0$. We can write by Newton's method for $n = 0$, and (A_2) that

$$\begin{aligned}
x^* - x_1 &= x^* - x_0 + F'(x_0)^{-1}F(x_0)\\
&= -F'(x_0)^{-1}\Big(F(x^*) - F(x_0) - F'(x_0)(x^* - x_0)\Big)\\
&= -F'(x_0)^{-1}\int_{x_0}^{x^*} F''(z)(x^* - z)dz\\
&= -F'(x_0)^{-1}\int_0^1 F''\Big(x_0 + \theta(x^* - x_0)\Big)(x^* - x_0)^2(1-\theta)d\theta\\
&= -F'(x_0)^{-1}\Big(\int_0^1 F''(x_0 + \theta(x^* - x_0)) - F''(x^*)\Big)(x^* - x_0)^2(1-\theta)d\theta\\
&\quad - \frac{1}{2}F'(x_0)^{-1}F''(x^*)(x^* - x_0)^2. \qquad (3.4)
\end{aligned}$$

Using $(A_2) - (A_4)$ and (3.4), we obtain in turn that

$$\begin{aligned}
\|x_1 - x^*\| &\leq \|F'(x_0)^{-1}F'(x^*)\|\|\int_0^1 F'(x^*)^{-1}\Big(F''(x_0 + \theta(x^* - x_0)) - F''(x^*))\Big)\times\\
&\quad (x_0 - x^*)^2(1-\theta)d\theta\| + \frac{1}{2}\|F'(x^0)^{-1}F'(x^*)\|\|F'(x^*)^{-1}F''(x^*)\|\|x_0 - x^*\|^2\\
&\leq \frac{\int_0^1 \overline{W}((1-\theta)\|x_0 - x^*\|)(1-\theta)d\theta + \frac{1}{2}\delta}{1 - p(\|x_0 - x^*\|)}\|x_0 - x^*\|^2\\
&\leq \|x_0 - x^*\| < r, \qquad (3.5)
\end{aligned}$$

which shows (3.1) for $n = 0$ and $x_1 \in U(x^*, r)$. The induction for (3.1) is completed if we replace x_0, x_1 by x_k, x_{k+1}, respectively in the preceding estimates. Then, from

$$\|x_{k+1} - x^*\| \leq q\|x_k - x^*\| < r,$$

where,

$$q = \frac{\int_0^1 \overline{W}\Big((1-\theta)|x_0 - x^*\|\Big)(1-\theta)d\theta + \frac{1}{2}\delta}{1 - p(\|x_0 - x^*\|)} \in [0, 1),$$

we deduce that

$$\lim_{k\to+\infty} x_k = x^*, \text{ and } x_{k+1} \in U(x^*, r).$$

Finally to show the uniqueness part, let $y^* \in \overline{U}(x^*, r)$ with $F(y^*) = 0$. As in (3.5) with x_0, x^* replaced by x_k, y^*, we arrive at

$$\begin{aligned}\|x_{k+1} - y^*\| &\leq \frac{\int_0^1 \overline{W}((1-\theta)\|x_k - y^*\|)(1-\theta)d\theta + \frac{1}{2}\delta}{1 - p(\|x_k - y^*\|)}\|x_k - y^*\|^2 \\ &\leq q^*\|x_k - y^*\|,\end{aligned}$$

where

$$q^* = \frac{\int_0^1 \overline{W}((1-\theta)\|x_0 - y^*\|)(1-\theta)d\theta + \frac{1}{2}\delta}{1 - p(\|x_0 - y^*\|)} \in [0, 1).$$

Hence, we get that

$$\lim_{k\to\infty} x_k = y^*$$

But, we know that

$$\lim_{k\to\infty} x_k = x^*$$

Therefore, we conclude that $x^* = y^*$.

□

Next, we present another local convergence result analogous to Theorem 3.1 But first, we need to introduce hypotheses (A^*):

(A_1^*) $F : D \subset X \to Y$ is a twice continuously Fréchet-differentiable operator.

(A_2^*) There exists $x^* \in D, \delta \geq 0$ such that $F(x^*) = 0, F'(x^*)^{-1} \in L(Y, X)$ and $\|F'(x^*)^{-1}F''(x^*)\| \leq \delta$.

(A_3^*) There exists a function $V : [0, \infty) \to [0, \infty)$ continuous and nondecreasing such that for each $x \in D$

$$\|F'(x^*)^{-1}\Big(F'(x) - F'(x^*)\Big)\| \leq V(\|x - x^*\|)$$

and $V(0) = 0$.
Assume equation V(t) = 1 has a minimal solution ξ_0.
Define the set

$$\Omega = D \cap U(x^*, \xi_0)$$

(A_4^*) There exists a function $W^* : [0, \xi_0) \to [0, \infty)$ continuous and nondecreasing such that for each $x \in \Omega$

$$\|F'(x^*)^{-1}\Big(F'(x) - F'(x^*)\Big)\| \leq W^*(\|x - x^*\|).$$

(A_5^*) $U(x^*, \xi) \subseteq D$
where the radius of convergence ξ is the smallest positive zero of function $h(t) = 0$ in the interval $(0, \xi_0)$,

$$h(t) = h_0(t) - 1$$

and

$$h_0(t) = \frac{\int_0^1 W^*\Big((1-\theta)t\Big)(1-\theta)d\theta + \frac{1}{2}\delta}{1 - V(t)} t.$$

Then, we can show the following local convergence result for Newton's method under the (A^*) conditions in an analogous to Theorem 3.1 way.

Theorem 3.2 Suppose that the (A^*) hypotheses are satisfied. Then, sequence $\{x_n\}$ generated for $x_0 \in U(x^*, \xi) - \{x^*\}$ by Newton's method is well defined in $U(x^*, \xi)$, remains in $U(x^*, \xi)$ for each $n = 0, 1, 2, \ldots$ and converges to x^* which is the only solution of equation $F(x) = 0$ in $U(x^*, \xi)$. Moreover, the following estimates hold

$$\|x_{n+1} - x^*\| \leq h_0(\|x_n - x^*\|)\|x_n - x^*\| \leq \|x_n - x^*\| < \xi. \tag{3.6}$$

Proof. Follow the proof of Theorem 3.1 but use the estimate

$$\|F'(x^*)^{-1}\Big(F'(x) - F'(x^*)\Big)\| \leq V(\|x - x^*\|) \leq V(r) < 1, \tag{3.7}$$

leading to

$$\|F'(x)^{-1}F'(x^*)\| \leq \frac{1}{1 - V(\|x - x^*\|)} \tag{3.8}$$

instead of (3.3). □

Concerning the uniqueness part of Theorem 3.2, we can have the following alternative which may be better.

Proposition 3.3 Suppose that the (A^*) are satisfied. Moreover, suppose that there exists $\xi^* \geq \xi$ such that

$$\int_0^1 V(\theta\xi^*)d\theta < 1, \tag{3.9}$$

then, the conclusions of Theorem 3.2 hold except the uniqueness part. Moreover, the point x^* is the only solution of equation $F(x) = 0$ in $B = D \cap U(x^*, \xi^*)$.

Proof. We only need to show the uniqueness part, since the rest of the proof is given in Theorem 3.2.
Let $y^* \in B$ with $F(y^*) = 0$. Let also

$$T = \int_0^1 F'(x^* + \theta(y^* - x^*))d\theta.$$

Using (A_3^*), we get that

$$\|F'(x^*)^{-1}\left(T - F'(x^*)\right)\| \leq \int_0^1 V(\theta\|g^* - x^*\|)d\theta \leq \int_0^1 V(\theta\xi^*)d\theta < 1,$$

so $T^{-1} \in L(Y, X)$. Then, from the identity

$$0 = F(y^*) - F(x^*) = T(y^* - x^*),$$

we conclude that $x^* = y^*$.

□

References

[1] Amat, S., Busquier, S. and Gutierrez, J.M., Geometric constructions of iterative functions to solve nonlinear equations, *J. Comput. Appl. Math.* 157, 2003, 197-205.

[2] Argyros, I.K., *Convergence and Applications of Newton-type Iterations*, Springer Verlag, Berlin, Germany, 2008.

[3] Argyros, I.K. and Magreñán, A.A., *A Contemporary Study of Iterative Methods, Academic Press*, Elsevier, San Diego, CA, USA, 2017.

[4] Denis, J.E. and Schnabel, R.B., *Numerical methods for unconstrained optimization and nonlinear equation*, SIAM, Philadelphia, 1996.

[5] Deulfhard, P. and Heindl, G., Affine invariant convergence theorems for Newton's method and extensions to related methods, *SIAM J. Numer. Anal.*, 16 (1979) 1 - 10.

[6] Ezquerro, J. A. and Hernández, M.A., A semi-local convergence result for Newton's method under generalized conditions of Kantorovich, *J. Complexity*, 30 (2014) 309 - 324.

[7] Ezquerro, J. A. and Hernández, M.A., Center conditions on high order derivates in the semi-local convergence of Newton's method, *J. Complexity*, 31 (2015) 277 - 292.

[8] Ezquerro, J. A. and Hernández M.A., Generalized differentiability conditions for Newton's method, *IMA J. Numer. Anal.*, 22, 2 (2002) 187 - 205.

[9] Ezquerro, J.A. and Hernández, M.A., Newton's method: an updated Approach of Kantorovich's theory, *Frontiers in Mathematics*, Birkhäuser, Cham Switzerland, (2017).

[10] Ezquerro, J. A. and Hernández M.A., On an application of Newton's method to non-linear operators with w-conditioned second derivative, *BIT*, 42, 3 (2002) 519 - 530.

[11] Ezquerro, J. A., González, D. and Hernández, M.A., Majorizing sequences for Newton's method from initial value problem, *J. Comput. Appl. Math.*, 236 (2012) 2246 - 2258.

[12] Gutiérrez, J. M., A new semi-local convergence theorem for Newton's method, *J. Comput. Appt. Math*, 79 (1997) 131-145.

[13] Gutiérrez, J. M. and Hernández, M.A., Majorizing sequences for Newton's method, *Tamkang J. Math.*, 29, 3 (1998) 199-202.

[14] Gutiérrez, J. M. and Hernández M.A., Newton's method under different Lipschitz conditions, (Numerical analysis and its applications) *Lectures Notes in Comput. Sci.*, 1988 (2000) 368 - 376.

[15] Gutiérrez, J. M. and Hernández, M.A., Newton's method under weak Kantorovich conditions, *IMA J. Numer. Anal.*, 20 (2000) 521 - 532.

[16] Hernández, M.A., Relaxing convergence conditions for Newton's method, *J. Math. Anal. Appl.*, 249 (2000) 463 - 475.

[17] Huang, Z., A note on the Kantorovich theorem for Newton method, *J. Comput. Appl. Math.*, 47 (1993) 211 - 217.

[18] Kantorovich, L.V. and Akilov, G. P., *Functional Analysis*, Pergamon Press, New York, 1982.

[19] Ortega, J. M. and Rheinboldt, W. C., *Iterative solutions of nonlinear equations in several variables*, Academic Press, New York, 1970.

Chapter 7

Extending the Applicability of Newton's Method under the k-th $(k \geq 2)$ Fréchet Derivaitive: Case II

1. Introduction

The semi-local convergence analysis of Newton's method given in [9, p. 110 - 120] is extended for Banach space valued operators, under the k-th $(k \geq 2)$ Fréchet-derivative, and without additional conditions.

2. Semi-Local Convergence Analysis

We first state the semi-local convergence analysis of Newton's method given in [9,17] in order to then compare with our new analysis that follows it. In particular, the comparison is given in Remark 2.3.

The semi-local convergence of Newton's method given in [9] is based on the hypotheses (C):

(C_1) $F : D \subseteq X \to Y$ is a nonlinear $k(k \geq 2)$ times continuously Fréchet-differentiable operator.

(C_2) There exist $x_0 \in D, \eta \geq 0$ and $b_i \geq 0, i = 2, 3, \ldots, k$ such that

$$F'(x_0)^{-1} \in L(Y, X), \|F'(x_0)^{-1}F(x_0)\| \leq \eta$$

and

$$\|F'(x_0)^{-1}F^{(i)}(x_0)\| \leq b_i.$$

(C_3) There exists function $\overline{W} : [0, \infty) \to \mathbb{R}$ such that

$$\|F'(x_0)^{-1}\left(F^{(k)}(x) - F^{(k)}(y)\right)\| \leq \overline{W}(\|x - y\|)$$

for each $x, y \in D$ with $\overline{W}(0) = 0$.

(C_4) Let function $\overline{g} : [t_0, \infty) \to \mathbb{R}$, where $t_0 \geq 0$ be defined by,

$$\overline{g}(t) = \int_{t_0}^{t}\int_{t_0}^{\theta_{k-1}} \cdots \int_{t_0}^{\theta_1} \overline{W}(\tau - t_0)d\tau d\theta_1 \ldots d\theta_{k-1} + \sum_{i=2}^{k} \frac{b_i}{i!}(t - t_0)^i - (t - t_0) + \eta,$$

$\overline{g}(\alpha) \leq 0$, α is the unique positive solution of $f'(t) = 0$, and $U(x_0, t^* - t_0) \subset D$, where $\overline{t}^*$ is the smallest positive solution of equation $\overline{g}(t) = 0$ in $[t_0, \infty)$.

Theorem 2.1 [9] Suppose that the hypotheses (C) are satisfied. Then, the sequence $\{x_n\}$ starting from x_0, generated by Newton's method, is well defined in $U(x_0, \overline{t}^* - t_0)$, remains in $U(x_0, \overline{t}^* - t_0)$ for each $n = 0, 1, 2, \ldots$, and converges to a solution $x^* \in U(x_0, \overline{t}^* - t_0)$ of equation $F(x) = 0$. Moreover, the following estimates hold

$$\|x_{n+1} - x_n\| \leq u_{n+1} - u_n$$

and

$$\|x_n - x^*\| \leq \overline{t}^* - u_n,$$

where

$$u_n = u_{n-1} - \frac{\overline{g}(\overline{u}_{n-1})}{\overline{g}'(u_{n-1})}.$$

Furthermore, the solution x^* is unique in $U(x_0, \overline{t}^{**} - t_0) \cap D$, if $\overline{t}^* < \overline{t}^{**}$ or in $U(x_0, \overline{t}^* - t_0)$, if $\overline{t}^* = \overline{t}^{**}$, provided that function $\overline{g}$ has two real roots $\overline{t}^*$ and $\overline{t}^{**}$

such that $t_0 < \bar{t}^* \leq \bar{t}^{**}$.

Our semi-local convergence of Newton's method is based on the hypotheses (H):

(H_1) = (C_1).

(H_2) = (C_2).

(H_3) There exists a function $V : [0,\infty) \to [0,\infty)$ continuous and nondecreasing such that for each $x \in D$

$$\|F'(x_0)^{-1}\Big(F'(x) - F'(x_0)\Big)\| \leq V(\|x - x_0\|)$$

and $V(0) = 0$.
Let $t_0 \geq 0$. Assume equation $V(t) = 1$ has a minimal positive solution r_0. Define the set D_0 by

$$D_0 = D \cap U(x_0, r_0).$$

(H_4) There exists a function $W : [0, r_0) \to [0,\infty)$ continuous and nondecreasing such that for each $x, y \in D_0$

$$\|F'(x_0)^{-1}\Big(F^{(k)}(x) - F^{(k)}(y)\Big)\| \leq W(\|x - y\|)$$

and $W(0) = 0$.
Let $t_0 \geq 0$. Define function $g : [t_0,\infty) \to \mathbb{R}$ by

$$g(t) = \int_{t_0}^{t}\int_{t_0}^{\theta_{k-1}} \cdots \int_{t_0}^{\theta_1} W(\tau - t_0)d\tau d\theta_1 \ldots d\theta_{k-1} + \frac{\delta}{2}(t-t_0)^2 + \sum_{i=2}^{k}\frac{b_i}{i!}(t-t_0)^i - (t-t_0) + \eta.$$

(H_5) $g(\gamma) \leq 0$ and γ is the unique positive solution of the equation $g'(t) = 0$ such that $\gamma \geq t_0$ and $U(x_0, t^* - t_0) \subset D$, where t^* is the smallest positive solution of equation $g(t) = 0$ in $[t_0,\infty)$.

(H_6) $V(t) \leq g'(t) + 1$ for each $t \in [t_0, \gamma]$.

(H_7) There exists $\gamma_1 \geq \gamma$ such that

$$\int_0^1 V\Big((1-\theta)\gamma+\theta\gamma_1\Big)d\theta < 1.$$

Set $D_1 = D \cap \overline{U}(x_0, \gamma_1)$.

Next, we present the semi-local convergence analysis of Newton's method using the preceding hypotheses (H) and notation.

Theorem 2.2 Suppose that the hypotheses (H) are satisfied. Then, sequence $\{x_n\}$ generated for $x_0 \in D$ by Newton's method is well defined in $\overline{U}(x_0, t^* - t_0)$, remains in $\overline{U}(x_0, t^* - t_0)$ for each $n = 0, 1, 2, \ldots$, and converges to a solution $x^* \in \overline{U}(x_0, t^* - t_0)$ of equation $F(x) = 0$. Moreover, the following estimates hold for each $n = 0, 1, 2, \ldots$

$$\|x_{n+1} - x_n\| \leq t_{n+1} - t_n$$

and

$$\|x_n - x^*\| \leq t^* - t_n,$$

where

$$t_n = t_{n-1} - \frac{g(t_{n-1})}{g'(t_{n-1})}.$$

Furthermore, the point x^* is the only solution of equation $F(x) = 0$ in D_1.

Proof. Let $x \in U(x_0, r_0)$. Using (H_3), we have that

$$\|F'(x_0)^{-1}\Big(F'(x) - F'(x_0)\Big)\| \leq V(\|x - x_0\|) \leq V(r_0) < 1. \tag{2.1}$$

Then, it follows from (2.1) and the Banach lemma on invertible operators [18] that $F'(x)^{-1} \in L(Y, X)$, and

$$\|F'(x)^{-1}F'(x_0)\| \leq \frac{1}{1 - V(\|x - x_0\|)}. \tag{2.2}$$

By (H_6) and (2.2) we get that

$$\|F'(x)^{-1}F'(x_0)\| \leq \frac{1}{1 - V(\|x - x_0\|)} \leq -\frac{1}{g'(\|x - x_0\|)}. \tag{2.3}$$

The rest of the proof until the uniqueness part follows exactly as in the proof given in [9,13,17], and by simply noticing that the iterates $\{x_n\}$ remains in D_0 which is at least as precise domain as D, since $D_0 \subseteq D$. To show the uniqueness part, let $y^* \in D_1$ with $F(y^*) = 0$. Define operator M by

$$M = \int_0^1 F'\Big(x^* + \theta(y^* - x^*)\Big)d\theta.$$

Using (H_3) and (H_7), we obtain in turn that

$$\begin{aligned} \|F'(x_0)^{-1}\Big(M - F'(x_0)\Big)\| &\leq \int_0^1 v\Big(\|x^* + \theta(y^* - x^*) - x_0\|\Big)d\theta \\ &\leq \int_0^1 v\Big((1-\theta)\|x^* - x_0\| + \theta\|y^* - x_0\|\Big) \\ &\leq \int_0^1 v\Big((1-\theta)\gamma + \theta\gamma_1\Big)d\theta < 1, \end{aligned}$$

so $M^{-1} \in L(Y,X)$. Finally, using the identity

$$0 = F(y^*) - F(x^*) = M(y^* - x^*),$$

we deduce that $x^* = y^*$.

□

Remark 2.3 (i) Suppose that hypotheses of Theorem 2.1 and Theorem 2.2 hold. Then, we have for each $t \in [t_0, \alpha)$

$$W(t) \leq \overline{W}(t), \tag{2.4}$$

$$g(t) \leq \overline{g}(t), \tag{2.5}$$

and

$$g'(t) \leq \overline{g}'(t). \tag{2.6}$$

If for each $t_0 < s \leq t$

$$-\frac{g(s)}{g'(s)} \leq -\frac{\overline{g}(t)}{\overline{g}'(t)}$$

and

$$s - \frac{g(s)}{g'(s)} \leq t - \frac{\overline{g}(t)}{\overline{g}'(t)},$$

then, a simple inductive argument shows that

$$t_n \leq u_n,$$

$$t_{n+1} - t_n \leq u_{n+1} - u_n,$$

so,

$$t^* \leq \overline{t}^*.$$

Hence, the error bounds on the distances are at least as tight and the information on the location of the solution x^* is at least as precise.

(ii) Clearly, in view of (2.4) - (2.6), the hypotheses of Theorem 2.1 imply the hypotheses of Theorem 2.2 but not necessarily vice versa.
For example from (2.6)

$$g(\alpha) \leq \overline{g}(\alpha) \leq 0,$$

the existence of γ is implied.

(iii) The results obtained here can be improved even further, if we replace the set D_0 by the set $\tilde{D_0}$ defined by

$$\tilde{D_0} = D \cap U(x_1, r_0 - \eta)$$

provided that $\eta < r_0$.

Notice that the iterated $\{x_n\}$ lie in $\tilde{D_0}$, which is a more precise location than D_0, since $\tilde{D_0} \subseteq D_0$. The new function $\tilde{W}$ replacing W in hypotheses (H) and Theorem 2.2 will be at least as tight as W leading to the least as weak sufficient semi-local convergence conditions for Newton's method. It is worth noticing that the advantages $(i) - (iii)$ are obtained under the same computational cost as in Theorem 2.1, since the computation of Lipschitz-type condition (C_3) passes through the computation of conditions (H_3), and condition (H_4) as special cases.

References

[1] Amat, S., Busquier, S. and Gutierrez, J.M., Geometric constructions of iterative functions to solve nonlinear equations, *J. Comput. Appl. Math.* 157, 2003, 197-205.

[2] Argyros, I.K., *Convergence and Applications of Newton-type Iterations*, Springer Verlag, Berlin, Germany, 2008.

[3] Argyros, I.K. and Magreñán, A.A., *A Contemporary Study of Iterative Methods, Academic Press*, Elsevier, San Diego, CA, USA, 2017.

[4] Denis, J.E. and Schnabel, R.B., *Numerical methods for unconstrained optimization and nonlinear equation*, SIAM, Philadelphia, 1996.

[5] Deulfhard, P. and Heindl, G., Affine invariant convergence theorems for Newton's method and extensions to related methods, *SIAM J. Numer. Anal.*, 16 (1979) 1 - 10.

[6] Ezquerro, J. A. and Hernández, M.A., A semi-local convergence result for Newton's method under generalized conditions of Kantorovich, *J. Complexity*, 30 (2014) 309 - 324.

[7] Ezquerro, J. A. and Hernández, M.A., Center conditions on high order derivates in the semi-local convergence of Newton's method, *J. Complexity*, 31 (2015) 277 - 292.

[8] Ezquerro, J. A. and Hernández M.A., Generalized differentiability conditions for Newton's method, *IMA J. Numer. Anal.*, 22, 2 (2002) 187 - 205.

[9] Ezquerro, J.A. and Hernández, M.A., Newton's method: an updated Approach of Kantorovich's theory, *Frontiers in Mathematics*, Birkhäuser, Cham Switzerland, (2017).

[10] Ezquerro, J. A. and Hernández M.A., On an application of Newton's method to non-linear operators with w-conditioned second derivative, *BIT*, 42, 3 (2002) 519 - 530.

[11] Ezquerro, J. A., González, D. and Hernández, M.A., Majorizing sequences for Newton's method from initial value problem, *J. Comput. Appl. Math.*, 236 (2012) 2246 - 2258.

[12] Gutiérrez, J. M., A new semi-local convergence theorem for Newton's method, *J. Comput. Appt. Math*, 79 (1997) 131-145.

[13] Gutiérrez, J. M. and Hernández, M.A., Majorizing sequences for Newton's method, *Tamkang J. Math.*, 29, 3 (1998) 199-202.

[14] Gutiérrez, J. M. and Hernández M.A., Newton's method under different Lipschitz conditions, (Numerical analysis and its applications) *Lectures Notes in Comput. Sci.*, 1988 (2000) 368 - 376.

[15] Gutiérrez, J. M. and Hernández, M.A., Newton's method under weak Kantorovich conditions, *IMA J. Numer. Anal.*, 20 (2000) 521 - 532.

[16] Hernández, M.A., Relaxing convergence conditions for Newton's method, *J. Math. Anal. Appl.*, 249 (2000) 463 - 475.

[17] Huang, Z., A note on the Kantorovich theorem for Newton method, *J. Comput. Appl. Math.*, 47 (1993) 211 - 217.

[18] Kantorovich, L.V. and Akilov, G. P., *Functional Analysis*, Pergamon Press, New York, 1982.

[19] Ortega, J. M. and Rheinboldt, W. C., *Iterative solutions of nonlinear equations in several variables*, Academic Press, New York, 1970.

Chapter 8

Convergence under w-Conditions and the k-Fréchet Derivatives: Case III

1. Introduction

The semi-local as well as the local convergence analysis of Newton's method for Banach space valued operators given in [9, p. 70, 76] derivatives $(k \geq 3)$ is extended, and without addition conditions.

2. Semi-Local Convergence Analysis

We first state the semi-local convergence analysis of Newton's method given in [9] in order to then compare with our new analysis that follows it. In particular, the comparison is given in Remark 2.3.

The semi-local convergence of Newton's method given in [9] is based on the hypotheses (C):

(C_1) $F : D \subseteq X \to Y$ is a nonlinear $k(k \geq 3)$ times continuously Fréchet-differentiable operator.

(C_2) There exist $x_0 \in D, \eta \geq 0$ and $b_i \geq 0, i = 3, \ldots, k-1$ such that

$$F'(x_0)^{-1} \in L(Y,X), \|F'(x_0)^{-1}F(x_0)\| \leq \eta$$

and

$$\|F'(x_0)^{-1}F^{(i)}(x_0)\| \leq b_i.$$

(C_3) There exists function $\overline{W} : [0, +\infty) \to \mathbb{R}$ such that

$$\|F'(x_0)^{-1}F^{(k)}(x)\| \leq \overline{W}(\|x\|)$$

for each $x \in D$ with $\overline{W}(0) = 0$.

Let $\overline{W}(t; \|x_0\|, t_0) := \overline{W}(t - t_0 + \|x_0\|)$.
Then, the preceding estimates can also be written as

$$\|F'(x_0)^{-1}F^{(k)}(x)\| \leq \overline{W}(t; \|x_0\|, t_0).$$

(C_4) Let function $\overline{g} : [t_0, +\infty) \to \mathbb{R}$, where $t_0 \geq 0$ be defined by,

$$\overline{g}(t) = \int_{t_0}^{t} \int_{t_0}^{\theta_{k-1}} \cdots \int_{t_0}^{\theta_1} \overline{W}(\tau, \|x_0\|, t_0) d\tau d\theta_1 \ldots d\theta_{k-1} + \sum_{i=2}^{k} \frac{b_i}{i!}(t - t_0)^i - (t - t_0) + \eta,$$

$\overline{g}(\alpha) \leq 0$; α is the unique positive solution of $f'(t) = 0$, and $U(x_0, t^* - t_0) \subset D$, where $\overline{t}^*$ is the smallest positive solution of equation $\overline{g}(t) = 0$ in $[t_0, \infty)$.

Theorem 2.1 [9] Suppose that the hypotheses (C) are satisfied. Then, the sequence $\{x_n\}$ starting from x_0 and generated by Newton's method, is well defined in $U(x_0, \overline{t}^* - t_0)$, remains in $U(x_0, \overline{t}^* - t_0)$ for each $n = 0, 1, 2, \ldots$, and converges to a solution $x^* \in U(x_0, \overline{t}^* - t_0)$ of equation $F(x) = 0$. Moreover, the following estimates hold

$$\|x_{n+1} - x_n\| \leq u_{n+1} - u_n$$

and

$$\|x_n - x^*\| \leq \overline{t}^* - u_n,$$

where

$$u_n = u_{n-1} - \frac{\overline{g}(\overline{u}_{n-1})}{\overline{g}'(u_{n-1})}.$$

Furthermore, the solution x^* is unique in $U(x_0, \overline{t}^{**} - t_0) \cap D$, if $\overline{t}^* < \overline{t}^{**}$ or in $U(x_0, \overline{t}^* - t_0)$, if $\overline{t}^* = \overline{t}^{**}$, provided that function $\overline{g}$ has two real roots $\overline{t}^*$ and $\overline{t}^{**}$ such that $t_0 < \overline{t}^* \leq \overline{t}^{**}$.

The new semi-local convergence of Newton's method uses the hypotheses (H):

(H_1) $F : D \subset X \to Y$ is a k- times $(k \geq 3)$ continuously Fréchet-differentiable operator.

(H_2) There exists $x_0 \in D, b_i \geq 0, i - 3, \ldots, k-1$ and $\eta \geq 0$ such that

$$F'(x_0)^{-1} \in L(Y, X), \|F'(x_0)^{-1}\| \leq b_i,$$

and

$$\|F'(x_0)^{-1}F(x_0)\| \leq \eta.$$

(H_3) There exists function $V : [0, \infty) \to [0, \infty)$ continuous and nondecreasing such that for each $x \in D$

$$\|F'(x_0)^{-1}\Big(F'(x) - F'(x_0)\Big)\| \leq V(\|x - x_0\|)$$

and $V(0) = 0$.
Let $t_0 \geq 0$. Assume equation $V(t) = 1$ has a minimal positive solution r_0. Define the set D_0 by

$$D_0 = D \cap U(x_0, r_0).$$

(H_4) There exists a function $W : [0, r_0) \to [0, \infty)$ continuous, and nondecreasing with $W(0) = 0$ such that for each $x \in \overline{U}(x_0, r_0 - t_0)$

$$\|F'(x_0)^{-1}F^{(k)}(x)\| \leq W(\|x\|).$$

Let $\overline{W}(t; \|x_0\|, t_0) := W(t - t_0 + \|x_0\|)$.
Then, the preceding estimate can also be written as

$$\|F'(x_0)^{-1}F^{(k)}(x)\| \leq \overline{W}(t; \|x_0\|, t_0).$$

Define the function $g : [t_0, r_0) \to \mathbb{R}$ by

$$g(t) = \int_{t_0}^{t} \int_{t_0}^{\theta_{k-1}} .. \int_{t_0}^{\theta_1} \overline{W}(\tau; \|x_0\|, t_0) d\tau d\theta_1 \ldots d\theta_{k-1} + \frac{\delta}{2}(t - t_0)^2 + \sum_{i=2}^{k} \frac{b_i}{i!}(t - t_0)^i$$

$$-(t - t_0) + \eta.$$

(H_5) $g(\gamma) \leq 0$, γ is the unique positive solution of the equation $g'(t) = 0$ such that $\gamma \geq t_0$, and $U(x_0, t^* - t_0) \subset D$, where t^* is the smallest positive solution of equation $g(t) = 0$ in $[t_0, \infty)$.

(H_6) $V(t) \leq g'(t) + 1$ for each $t \in [t_0, \gamma]$.

(H_7) There exists $\gamma_1 \geq \gamma$ such that

$$\int_0^1 V\Big((1-\theta)\gamma + \theta\gamma_1\Big) d\theta < 1.$$

Set $D_1 = D \cap \overline{U}(x_0, \gamma_1)$.

Next, we present the semi-local convergence analysis of Newton's method using the preceding hypotheses (H) and notation.

Theorem 2.2 Suppose that the hypotheses (H) are satisfied. Then, sequence $\{x_n\}$ generated for $x_0 \in D$ by Newton's method is well defined in $\overline{U}(x_0, t^* - t_0)$, remains in $\overline{U}(x_0, t^* - t_0)$ for each $n = 0, 1, 2, \ldots$, and converges to a solution $x^* \in \overline{U}(x_0, t^* - t_0)$ of equation $F(x) = 0$. Moreover, the following estimates hold for each $n = 0, 1, 2, \ldots$

$$\|x_{n+1} - x_n\| \leq t_{n+1} - t_n$$

and

$$\|x_n - x^*\| \leq t^* - t_n,$$

where

$$t_n = t_{n-1} - \frac{g(t_{n-1})}{g'(t_{n-1})}.$$

Furthermore, the point x^* is the only solution of equation $F(x) = 0$ in D_1.

Proof. Let $x \in U(x_0, r_0)$. Using (H_3), we have that

$$\|F'(x_0)^{-1}\Big(F'(x) - F'(x_0)\Big)\| \leq V(\|x - x_0\|) \leq V(r_0) < 1. \tag{2.1}$$

Then, it follows from (2.1) and the Banach lemma on invertible operators [18] that $F'(x)^{-1} \in L(Y, X)$ and

$$\|F'(x)^{-1}F'(x_0)\| \leq \frac{1}{1 - V(\|x - x_0\|)}. \tag{2.2}$$

By (H_6) and (2.2), we get that

$$\|F'(x)^{-1}F'(x_0)\| \leq \frac{1}{1 - V(\|x - x_0\|)} \leq -\frac{1}{g'(\|x - x_0\|)}. \tag{2.3}$$

The rest of the proof until the uniqueness part follows exactly as in the proof given in [9], and by simply noticing that the iterates $\{x_n\}$ remains in D_0 which is at least s precise domain as D, since $D_0 \subseteq D$. To show the uniqueness part, let $y^* \in D_1$ with $F(y^*) = 0$. Define operator M by

$$M = \int_0^1 F'\Big(x^* + \theta(y^* - x^*)\Big) d\theta.$$

Using (H_3) and (H_7), we obtain in turn that

$$\begin{aligned}
\|F'(x_0)^{-1}\Big(M - F'(x_0)\Big)\| &\leq \int_0^1 V\Big(\|x^* + \theta(y^* - x^*) - x_0\|\Big) d\theta \\
&\leq \int_0^1 V\Big((1-\theta)\|x^* - x_0\| + \theta\|y^* - x_0\|\Big) \\
&\leq \int_0^1 V\Big((1-\theta)\gamma + \theta\gamma_1\Big) d\theta < 1,
\end{aligned}$$

so $M^{-1} \in L(Y, X)$. Finally, using the identity

$$0 = F(y^*) - F(x^*) = M(y^* - x^*),$$

we deduce that $x^* = y^*$.

□

Remark 2.3 (i) Suppose that hypotheses of Theorem 2.1 and Theorem 2.2 hold. Then, we have for each $t \in [t_0, \alpha)$

$$W(t) \leq \overline{W}(t), \tag{2.4}$$

$$g(t) \leq \overline{g}(t), \tag{2.5}$$

and

$$g'(t) \leq \overline{g}'(t). \tag{2.6}$$

If for each $t_0 < s \leq t$

$$-\frac{g(s)}{g'(s)} \leq -\frac{\overline{g}(t)}{\overline{g}'(t)}$$

and

$$s - \frac{g(s)}{g'(s)} \leq t - \frac{\overline{g}(t)}{\overline{g}'(t)},$$

then, a simple inductive argument shows that

$$t_n \leq u_n,$$

$$t_{n+1} - t_n \leq u_{n+1} - u_n,$$

so,

$$t^* \leq \overline{t}^*.$$

Hence, the error bounds on the distances are at least as tight, and the information on the location of the solution x^* is at least as precise.

(ii) Clearly, in view of (2.4) - (2.6), the hypotheses of Theorem 2.1 imply the hypotheses of Theorem 2.2 but not necessarily vice versa.
For example from (2.6)

$$g(\alpha) \leq \overline{g}(\alpha) \leq 0,$$

the existence of γ is implied.

(iii) The results obtained here can be improved even further, if we replace the set D_0 by the set $\tilde{D}_0$ defined by

$$\tilde{D}_0 = D \cap U(x_1, r_0 - \eta)$$

provided that $\eta < r_0$.

Notice that the iterated $\{x_n\}$ lie in $\tilde{D}_0$ which is a more precise location that D_0, since $\tilde{D}_0 \subseteq D_0$. The new function $\tilde{W}$ replacing W in hypotheses (H) and Theorem 2.2 will be at least as tight as W leading to the least as weak sufficient semi-local convergence conditions for Newton's method. It is worth noticing that the advantages $(i)-(iii)$ are obtained under the same computational cost as in Theorem 2.1, since the computation of Lipschitz-type condition (C_3) passes through the computation of conditions (H_3) and condition (H_4) as special cases.

3. Local Convergence Analysis

The local convergence analysis of Newton's method (not given in [9]) is based on hypotheses (A):

(A_1) $F : D \subset X \to Y$ is a nonlinear $k (k \geq 3)$ times continuously Fréchet-differentiable operator.

(A_2) There exists $x^* \in D$, and $b_i \geq 0, i = 3, 4, \ldots, k-1$ such that

$$F(x^*) = 0, F'(x^*)^{-1} \in L(Y, X),$$

and

$$\|F'(x^*)^{-1} F^{(i)}(x^*)\| \leq b_i.$$

(A_3) There exists a function $V : [0, \infty) \longrightarrow [0, \infty)$ continuous, and nondecreasing with $V(0) = 0$ such that for each $x \in D$

$$\|F'(x^*)^{-1}\Big(F'(x) - F'(x^*)\Big)\| \leq V(\|x - x^*\|).$$

Assume equation $V(t) = 1$ has a minimal positive solution r_0.
Set $D_0 = D \cap U(x^*, r_0)$.

(A_4) There exists a function $W : [0, r_0) \longrightarrow [0, \infty)$ continuous and nondecreasing with $W(0) = 0$ such that for each $x \in D$

$$\|F'(x^*)^{-1} F^{(k)}(x)\| \leq W(\|x - x^*\|).$$

Define functions h and h_0 on the interval $[0,r_0)$ by

$$h(t) = \frac{\sum_{i=2}^{k-1}\frac{1}{(i-1)!}b_i t^{i-1} + \frac{1}{(k-3)!}\int_0^1\int_0^1 W\Big(\theta(1-\tau)t\Big)(1-\theta)^{k-3}(1-\tau)t^{k-1}d\theta d\tau}{1-V(t)}$$

and

$$h_0(t) = h(t) - 1.$$

Notice that $h_0(0) = -1 < 0$ and $h_0(t) \to \infty$ as $t \to r_0^-$. Then, it follows from the intermediate value theorem that function h_0 has zeros in the interval $(0,r_0)$. Denote by r the smallest such zero.

(A_5) $\overline{U}(x^*,r) \subseteq D$.

(A_6) There exists $r^* \geq r$ such that

$$\int_0^1 V(\theta r^*)d\theta < 1.$$

Define the set $D_1 = D \cap \overline{U}(x^*,r^*)$.

Next, we can show the following local convergence analysis of Newton's method using the preceding notation and the hypotheses (A).

Theorem 3.1 Suppose that the hypotheses (A) are satisfied. Then, sequence $\{x_n\}$ generated for $x_0 \in U(x^*,r) - \{x^*\}$ by Newton's method is well defined in $U(x^*,r)$, remains in $U(x^*,r)$ for each $n = 0,1,2,\ldots$, and converges to x^* which is the only solution of equation $F(x) = 0$ in D_1. Moreover, the following estimates hold

$$\|x_{n+1} - x^*\| \leq h(\|x_n - x^*\|)\|x_n - x^*\| \leq \|x_n - x^*\| < r. \tag{3.1}$$

Proof. We shall use mathematical induction on m to show that estimates (3.1) holds. Let $x \in U(x^*,r_0)$ and $k = 3,4\ldots$. By Taylor series about x^* we obtain the following identity

$$F''(x) = \sum_{i=2}^{k-1} \frac{1}{(i-2)!} F^{(i)}(x^*)(x-x^*)^{i-2}$$
$$+ \frac{1}{(k-3)!} \int_0^1 \Big(F^{(k)}(x^* + \theta(x-x^*)) \Big) (x-x^*)^{k-2}(1-\theta)^{k-3} d\theta. \quad (3.2)$$

Using the definition of r_0 and hypothesis (A_3) we get that

$$\|F'(x^*)^{-1}\Big(F'(x) - F'(x^*)\Big)\| \le V(\|x-x^*\|) \le V(r) \le V(r_0) < 1. \quad (3.3)$$

It follows from (3.3) and the Banach lemma on invertible operators [18] that $F'(x)^{-1} \in L(Y,X)$, and

$$\|F'(x)^{-1}F'(x^*)\| \le \frac{1}{1 - V(\|x-x^*\|)}. \quad (3.4)$$

In particular for $x = x_0, x_1$ is well defined by Newton's method for $n = 0$. Suppose x_m are well defined of $n, m = 0, 1, 2, \ldots, n$. In view of Newton's method, we have the approximation

$$\begin{aligned} x^* - x_{m+1} &= x^* - x_m + F'(x_m)^{-1}F(x_m) \\ &= -F'(x_m)^{-1}\Big[F(x^*) - F(x_m) - F'(x_m)(x^* - x_m)\Big] \\ &= -F'(x_m)^{-1} \int_{x_m}^{x^*} F''(z)(x^* - z) dz \\ &= -F'(x_m)^{-1}F'(x^*) \times Al, \end{aligned} \quad (3.5)$$

so

$$\|x_{m+1} - x^*\| \le \|Al\|, \quad (3.6)$$

where

$$Al = \int_0^1 F'(x^*)^{-1}F''\Big(x_m + \tau(x^* - x_m)\Big)(x^* - x_m)^2(1-\tau)d\tau.$$

We need an upper bound on $\|F'(x^*)^{-1}F''(x)\|$. To find such a bound we use hypotheses $(A_2),(A_4)$ and (3.2) to obtain in turn that.

$$\begin{aligned}\|F'(x^*)^{-1}F''(x)\| &\leq \sum_{i=2}^{k-1}\frac{1}{(i-2)!}\|F'(x^*)^{-1}F^{(i)}(x^*)\|\|x-x^*\|^{i-2}\\ &+\frac{1}{(k-3)!}\int_0^1\|F'(x^*)^{-1}\Big(F^{(k)}(x^*+\theta(x-x^*))\Big)-F^{(k)}(x^*)\|\|x-x^*\|^{k-2}(1-\theta)^{k-3}d\theta\\ &\leq \sum_{i=2}^{k-1}\frac{1}{(i-2)!}b_i\|x-x^*\|^{i-2}+\frac{1}{(k-3)!}\int_0^1 W(\theta\|x-x^*\|)\|x-x^*\|^{k-2}(1-\theta)^{k-3}d\theta.\end{aligned} \tag{3.7}$$

In particular, for $x=x_m+\tau(x^*-x_m)$, we obtain from (3.7) that

$$\begin{aligned}&\|F'(x^*)^{-1}F''(x_m+\tau(x^*-x_m))\|\\ &\leq \sum_{i=2}^{k-1}\frac{1}{(i-2)!}b_i\int_0^1(1-\tau)^{i-2}\|x^*-x_m\|^{i-2}\|x^*-x_m\|^2d\tau\\ &+\frac{1}{(k-3)!}\int_0^1\int_0^1 W\Big(\theta(1-\tau)\|x^*-x_m\|\Big)(1-\tau)^{k-2}\|x^*-x_m\|^{k-2}\|x^*-x_m\|^2(1-\theta)^{k-3}d\theta d\tau\\ &\leq \sum_{i=2}^{k-1}\frac{1}{(i-1)!}b_i\|x^*-x_m\|^i\\ &+\frac{1}{(k-3)!}\int_0^1\int_0^1 W(\theta(1-\tau)\|x^*-x_m\|)(1-\tau)^{k-2}(1-\theta)^{k-3}d\theta d\tau\|x^*-x_m\|^k.\end{aligned} \tag{3.8}$$

By combining (3.4) and (3.8) in (3.6) and using the definition of r and function h we get in turn that

$$\|x_{m+1}-x^*\|\leq h(\|x^*-x_m\|)\|x^*-x_m\|\leq\|x^*-x_m\|<r, \tag{3.9}$$

which completes the induction for (3.1).
Then, from the estimate

$$\|x_{m+1}-x^*\|\leq q\|x_m-x^*\|<r, \tag{3.10}$$

where $q=h(\|x^*-x_0\|)\in[0,1)$, we deduce that $\lim_{m\to\infty}=x^*$, and $x_{m+1}\in U(x^*,r)$.

Finally, to show the uniqueness part, let $y^* \in D_1$ with $F(y^*) = 0$. Define the linear operator

$$T = \int_0^1 F\Big(x^* + \theta(y^* - x^*)\Big)d\theta.$$

Using (A_3) we get that

$$\|F'(x^*)^{-1}\Big(T - F'(x^*)\Big)\| \leq \int_0^1 V(\theta\|y^* - x^*\|)d\theta \int_0^1 V(\theta r^*)d\theta < 1, \tag{3.11}$$

so $T^{-1} \in L(Y,X)$. Then, from the approximation

$$0 = F(y^*) - F(x^*) = T(y^* - x^*), \tag{3.12}$$

we conclude that $x^* = y^*$.

□

The case $k = 2$ is studied next, using the hypotheses (B):

(B_1) $F : D \subset X \to Y$ is a twice continuously Fréchet-differentiable operator.

(B_2) There exists $x^* \in D$, and $b_2 \geq 0$ such that

$$F(x^*) = 0, F'(x^*)^{-1} \in L(Y,X)$$

and

$$\|F'(x^*)^{-1}F''(x^*)\| \leq b_2.$$

(B_3) $= (A_3)$

(B_4) $= (A_4)$ for $k = 2$. Define functions h and h_0 on the interval $[0, r_0)$ by

$$h(t) = \frac{\int_0^1 W\Big((1-\theta)t\Big)d\theta + \frac{1}{2}b_2}{1 - V(t)}t,$$

and

$$h_0(t) = h(t) - 1.$$

Denote again by ρ the smallest zero of function h in the interval $(0, r_0)$.

(B_5) $\overline{U}(x^*,\rho) \subseteq D$.

(B_6) There exists $\rho^* \geq \rho$ such that

$$\int_0^1 V(\theta\rho^*)d\theta < 1.$$

Define the set $D_1 = D \cap U(x^*,\rho^*)$.

Using the preceding notation, we arrive at:

Theorem 3.2 Suppose that the hypotheses (B) we satisfied. Then, the conclusions of Theorem 3.1 hold.

Proof. Notice that we obtain the numerator from the definition of function h as follows:

$$\begin{aligned}
&\| \int_0^1 F'(x^*)^{-1}\Big(F''\big(x_m+\tau(x^*-x_m)\big)\Big)(1-\tau)(x^*-x_m)^2 d\tau\| \\
&\leq \| \int_0^1 F'(x^*)^{-1}\Big(F''\big(x_m+\tau(x^*-x_m)\big)-F''(x^*)\Big)(1-\tau)d\tau\| \, \|x^*-x_m\|^2 \\
&\qquad + \int_0^1 (1-\tau)d\tau \|F'(x^*)^{-1}F''(x^*)\| \, \|x^*-x_m\|^2 \\
&\leq \int_0^1 W\Big((1-\theta)\|x^*-x_m\|\Big)(1-\tau)d\tau\|x^*-x_m\|^2 + \frac{1}{2}b_2\|x^*-x_m\|^2. \qquad (3.13)
\end{aligned}$$

□

Remark 3.3 In view of the estimate

$$\begin{aligned}
&F'(x^*)^{-1}\Big(F'(x^*)-F(x)\Big) \\
&= -F'(x^*)^{-1}\Big[F'(x)-F'(x^*)-F''(x^*)(x-x^*)+F''(x^*)(x-x^*)\Big] \\
&= -F'(x^*)^{-1}\int_0^1 \Big(F''(x^*+\tau(x-x^*))-F''(x^*)\Big)d\tau(x-x^*) - F'(x^*)^{-1}F''(x^*)(x-x^*), \quad (3.14)
\end{aligned}$$

we can obtain by hypotheses (B_2) and (B_4) that

$$\begin{aligned}\|F'(x^*)^{-1}\big(F'(x)-F'(x^*)\big)\| &\leq \|\int_0^1 F'(x^*)^{-1}\big(F''(x^*+\tau(x-x^*))-F''(x^*)\big)d\tau(x-x^*)\| \\ &\quad +\|F'(x^*)^{-1}F''(x^*)(x-x^*)\| \\ &\leq \int_0^1 W(\tau\|x-x^*\|)d\tau\|x-x^*\|+b_2\|x-x^*\|. \end{aligned} \tag{3.15}$$

Hence, we can choose

$$V(t)=\Big(\int_0^1 W(\tau t)d\tau+b_2\Big)t. \tag{3.16}$$

However, the function V can be smaller and more flexible than its special choice given by (3.16) leading to tighter error bounds, and a larger radius of convergence.

References

[1] Amat, S., Busquier, S. and Gutierrez, J.M., Geometric constructions of iterative functions to solve nonlinear equations, *J. Comput. Appl. Math.* 157, 2003, 197-205.

[2] Argyros, I.K., *Convergence and Applications of Newton-type Iterations*, Springer Verlag, Berlin, Germany, 2008.

[3] Argyros, I.K. and Magreñán, A.A., *A Contemporary Study of Iterative Methods*, Academic Press, Elsevier, San Diego, CA, USA, 2017.

[4] Denis, J.E. and Schnabel, R.B., *Numerical methods for unconstrained optimization and nonlinear equation*, SIAM, Philadelphia, 1996.

[5] Deulfhard, P. and Heindl, G., Affine invariant convergence theorems for Newton's method and extensions to related methods, *SIAM J. Numer. Anal.*, 16 (1979) 1 - 10.

[6] Ezquerro, J. A. and Hernández, M.A., A semi-local convergence result for Newton's method under generalized conditions of Kantorovich, *J. Complexity*, 30 (2014) 309 - 324.

[7] Ezquerro, J. A. and Hernández, M.A., Center conditions on high order derivates in the semi-local convergence of Newton's method, *J. Complexity*, 31 (2015) 277 - 292.

[8] Ezquerro, J. A. and Hernández M.A., Generalized differentiability conditions for Newton's method, *IMA J. Numer. Anal.*, 22, 2 (2002) 187 - 205.

[9] Ezquerro, J.A. and Hernández, M.A., Newton's method: an updated Approach of Kantorovich's theory, *Frontiers in Mathematics*, Birkhäuser, Cham Switzerland, (2017).

[10] Ezquerro, J. A. and Hernández M.A., On an application of Newton's method to non-linear operators with w-conditioned second derivative, *BIT*, 42, 3 (2002) 519 - 530.

[11] Ezquerro, J. A., González, D. and Hernández, M.A., Majorizing sequences for Newton's method from initial value problem, *J. Comput. Appl. Math.*, 236 (2012) 2246 - 2258.

[12] Gutiérrez, J. M., A new semi-local convergence theorem for Newton's method, *J. Comput. Appt. Math*, 79 (1997) 131-145.

[13] Gutiérrez, J. M. and Hernández, M.A., Majorizing sequences for Newton's method, *Tamkang J. Math.*, 29, 3 (1998) 199-202.

[14] Gutiérrez, J. M. and Hernández M.A., Newton's method under different Lipschitz conditions, (Numerical analysis and its applications) *Lectures Notes in Comput. Sci.*, 1988 (2000) 368 - 376.

[15] Gutiérrez, J. M. and Hernández, M.A., Newton's method under weak Kantorovich conditions, *IMA J. Numer. Anal.*, 20 (2000) 521 - 532.

[16] Hernández, M.A., Relaxing convergence conditions for Newton's method, *J. Math. Anal. Appl.*, 249 (2000) 463 - 475.

[17] Huang, Z., A note on the Kantorovich theorem for Newton method, *J. Comput. Appl. Math.*, 47 (1993) 211 - 217.

[18] Kantorovich, L.V. and Akilov, G. P., *Functional Analysis*, Pergamon Press, New York, 1982.

[19] Ortega, J. M. and Rheinboldt, W. C., *Iterative solutions of nonlinear equations in several variables*, Academic Press, New York, 1970.

Chapter 9

Convergence under w-Conditions on the Second Derivative: Case IV

1. Introduction

The applicability of Newton's method given in [9, p.70] for w-bounded second Fréchet derivative, and Banach space value operator is extended without additional conditions.

2. Semi-Local Convergence Analysis

We first state the semi-local convergence analysis of Newton's method given in [9, p. 70] in order to then compare with our new analysis that follows it. In particular, the comparison is given in Remark 2.3.

The semi-local convergence of Newton's method given in [9] is presented under the hypotheses (C):

(C_1) $F : D \subset X \to Y$ is a twice continuously Fréchet-differentiable operator.

(C_2) There exist $x_0 \in D, \delta \geq 0$ and $\eta \geq 0$ such that

$$F'(x_0) \in L(Y,X), \|F'(x_0)^{-1}F(x_0)\| \leq \eta$$

and

$$\|F'(x_0)^{-1}F''(x_0)\| \leq \delta.$$

(C_3) There exists a function $\overline{W} : [0,\infty) \to [0,\infty)$ continuous, and nondecreasing such that for each $x \in D$

$$\|F'(x_0)^{-1}F''(x)\| \leq \overline{W}(\|x\|) \leq \tilde{\overline{W}}(t;\|x_0\|,t_0)$$

and $\overline{W}(0) = 0$,
where

$$\tilde{\overline{W}} = \begin{cases} \overline{W}(\|x_0\| - t_0 + t), \ if\ \overline{W} \text{ is nondecreasing provided that} \\ \qquad \|x\| - \|x_0\| \leq \|x - x_0\| \leq t - t_0 \\ \overline{W}(\|x_0\| + t_0 - t), \ if\ \overline{W} \text{ is nondecreasing provided that} \\ \qquad \|x\| - \|x_0\| \leq \|x - x_0\| \leq t - t_0. \end{cases}$$

Let $t_0 \geq 0$. Define function $\overline{g} : [t_0,\infty) \to \mathbb{R}$ by

$$\overline{g}(t) = \int_{t_0}^{t}\int_{t_0}^{\theta} \overline{W}(\tau;\|x_0\|,t_0)d\tau d\theta - (t - t_0) + \eta.$$

(C_4) $\overline{g}(\alpha) \leq 0$, α is the unique positive solution of the equation $\overline{g}\,'(t) = 0$ such that $\alpha > t_0$ and $U(x_0, t^* - t_0) \subseteq D$, where $\bar{t}^{\,*}$ is the smallest positive solution of equation $\overline{g}(t) = 0$ in $[t_0,\infty)$.

Theorem 2.1 [9] Suppose that the hypotheses (C) are satisfied. Then, sequence $\{x_n\}$ generated for $x_0 \in D$ by Newton's method is well defined in $\overline{U}(x_0,\bar{t}^*)$, remains in $\overline{U}(x_0,\bar{t}^*)$ for each $n = 0,1,2,\ldots$, and converges to a solution $x^* \in \overline{U}(x_0,\bar{t}^*)$ of equation $F(x) = 0$.
Moreover, the following estimates hold for each $n = 0,1,2,\ldots$

$$\|x_{n+1} - x_n\| \leq u_{n+1} - u_n$$

and

$$\|x_n - x^*\| \leq \bar{t}^{\,*} - u_n,$$

where

$$u_n = u_{n-1} - \frac{\overline{g}(u_{n-1})}{\overline{g}\,'(u_{n-1})}.$$

Furthermore, the solution x^* is unique in $U(x_0, \overline{t}^{**} - t_0) \cap D$, if $\overline{t}^* < \overline{t}^{**}$ or in $U(x_0, \overline{t}^* - t_0)$, if $\overline{t}^* = \overline{t}^{**}$, provided that function $\overline{g}$ has two real zeros $\overline{t}^*$ and $\overline{t}^{**}$ such that $t_0 < \overline{t}^* \leq \overline{t}^{**}$.

Our semi-local convergence of Newton's method is based on the hypothesis (H):

(H_1) $F : D \subset X \to Y$ is a twice continuously Fréchet-differentiable operator.

(H_2) There exist $x_0 \in D, \delta \geq 0$ and $\eta \geq 0$ such that

$$F'(x_0)^{-1} \in L(Y,X), \|F'(x_0)^{-1}F(x_0)\| \leq \eta$$

and

$$\|F'(x_0)^{-1}F''(x_0)\| \leq \delta.$$

(H_3) There exists a function $V : [0,\infty) \to [0,\infty)$ continuous, and nondecreasing such that for each $x \in D$

$$\|F'(x_0)^{-1}\Big(F(x) - F(x_0)\Big)\| \leq V(\|x - x_0\|)$$

and

$$V(0) = 0.$$

Let $t_0 \geq 0$. Assume equation $V(t) = 1$ has a minimal positive solution r_0, and let D_0 be

$$D_0 = D \cap U(x_0, r_0).$$

(H_4) There exists a function $W : [t_0, r_0) \to [0,\infty)$ continuous, and nondecreasing such that for each $x \in D_0$

$$\|F'(x_0)^{-1}F''(x)\| \leq W(\|x - x_0\|)$$

and

$$W(0) = 0.$$

Let $t_0 \geq 0$. Define function $g : [t_0, \infty) \to \mathbb{R}$ by

$$g(t) = \int_{t_0}^{t} \int_{t_0}^{\theta} \overline{W}(\tau; \|x_0\|, t_0) d\tau d\theta - (t - t_0) + \eta.$$

(H_5) $g(\gamma) \leq 0$, γ is the unique positive solution of the equation $g'(t) = 0$ such that $\gamma \geq t_0$ and $U(x_0, t^* - t_0) \subset D$, where t^* is the smallest positive solution of equation $g(t) = 0$ in $[t_0, \infty)$.

(H_6) $V(t) \leq g'(t) + 1$ for each $t \in [t_0, \gamma]$.

(H_7) There exists $\gamma_1 \geq \gamma$ such that

$$\int_0^1 V((1-\theta)\gamma + \theta\gamma_1) d\theta < 1.$$

Set $D_1 = D \cap \overline{U}(x_0, \gamma_1)$.

Next, we present the semi-local convergence analysis of Newton's method using the preceding notation and hypotheses (H).

Theorem 2.2 Suppose that the hypothesis (H) are satisfied. Then, sequence $\{x_n\}$ generated for $x_0 \in D$ by Newton's method is well defined in $\overline{U}(x_0, t^* - t_0)$, remains in $\overline{U}(x_0, t^*)$ for each $n = 0, 1, 2, \ldots$, and converges to a solution $x^* \in \overline{U}(x_0, t^* - t_0)$ of equation $F(x) = 0$.
Moreover, the following estimates holds for each $n = 0, 1, 2, \ldots$

$$\|x_{n+1} - x_n\| \leq t_{n+1} - t_n$$

and

$$\|x_n - x^*\| \leq t^* - t_n,$$

where

$$t_n = t_{n-1} - \frac{g(t_{n-1})}{g'(t_{n-1})}.$$

Furthermore, the point x^* is the only solution of equation $F(x) = 0$ in D_1.

Proof. Let $x \in U(x_0, r_0)$. Using (H_3), we have that

$$\|F'(x_0)^{-1}\Big(F'(x) - F'(x_0)\Big)\| \leq V(\|x - x_0\|) \leq V(r_0) < 1. \tag{2.1}$$

Then, it follows from (2.1) and the Banach Lemma on invertible operators [18] that $F'(x)^{-1} \in L(Y,X)$, and

$$\|F'(x)^{-1}F'(x_0)\| \leq \frac{1}{1 - V(\|x - x_0\|)}. \tag{2.2}$$

By (H_6) and (2.2) we get that

$$\|F'(x)^{-1}F'(x_0)\| \leq \frac{1}{1 - V(\|x - x_0\|)} \leq -\frac{1}{g'(\|x - x_0\|)}. \tag{2.3}$$

The rest of the proof until the uniqueness part follows exactly as in the proof given in [9]. and by simply noticing that the iterates $\{x_n\}$ remain in D_0 which is an at least as precise domain as D, since $D_0 \subseteq D$. To show the uniqueness part, let $y^* \in D_1$ with $F(y^*) = 0$. Define operator M by

$$M = \int_0^1 F'(x^* + \theta(y^* - x^*))d\theta.$$

Using (H_3) and (H_7) we obtain in turn that

$$\begin{aligned}
\|F'(x_0)^{-1}\Big(M - F'(x_0)\Big)\| &\leq \int_0^1 V\Big(\|x^* + \theta(y^* - x^*) - x_0\|\Big)d\theta \\
&\leq \int_0^1 V\Big((1-\theta)\|x^* - x_0\| + \theta\|y^* - x_0\|\Big) \\
&\leq \int_0^1 V\Big((1-\theta)\gamma + \theta\gamma_1\Big)d\theta < 1,
\end{aligned}$$

so $M^{-1} \in L(Y,X)$. Finally, using the identity

$$0 = F(y^*) - F(x^*) = M(y^* - x^*),$$

we deduce that $x^* = y^*$. □

Remark 2.3 **(i)** Suppose that the hypotheses of Theorem 2.1 and Theorem 2.3 hold. Then, we have for each $t \in [t_0, \alpha)$

$$W(t) \leq \overline{W}(t), \tag{2.4}$$

$$g(t) \leq \overline{g}(t) \tag{2.5}$$

$$g'(t) \leq \overline{g}\,'(t). \tag{2.6}$$

Then, if for each $t_0 < s \leq t$

$$-\frac{g(s)}{g'(s)} \leq -\frac{\overline{g}(t)}{\overline{g}'(t)}$$

and

$$s - \frac{g(s)}{g'(s)} \leq t - \frac{\overline{g}(t)}{\overline{g}'(t)},$$

a simple inductive argument shows that

$$t_n \leq u_n$$

$$t_{n+1} - t_n \leq u_{n+1} - u_n$$

so,

$$t^* \leq \overline{t}^*.$$

Hence, the error bounds on the distances are at least as tight, and the information on the location of the solution x^* is at least as precise.

(ii) Clearly, in view of (2.4) - (2.6), the hypotheses of Theorem 2.1 imply the hypotheses of Theorem 2.2 but not necessarily the vice versa.
For example from (2.6)

$$g(\alpha) \leq \overline{g}(\alpha) \leq 0,$$

the existence of the γ is implied.

(iii) The results obtained here can be improved even further, if we replace the set D_0 by the set $\tilde{D}_0$ defined by

$$\tilde{D}_0 = D \cap U(x_1, r_0 - \eta)$$

provided that $\eta < r_0$.

Notice that the iterates $\{x_n\}$ lie in $\tilde{D_0}$ which is a more precise location than D_0, since $\tilde{D_0} \subseteq D_0$. The new function $\tilde{W}$ replacing W in hypothesis (H) and Theorem 2.2 will be at least as tight as W leading to at least as weak sufficient semi-local convergence conditions for Newton's method. It is worth noticing that the advantages (i) - (iii) are obtained under the same computational cost as in Theorem 2.1, since the computation of the Lipschitz-type condition (C_3) passes through the computation of condition (H_3), and condition (H_4) as special cases.

3. Local Convergence Analysis

We first base the local convergence analysis (not given in [9]) on the hypotheses (A):

(A_1) $F : D \subset X \to Y$ is a twice continuously Fréchet-differentiable operator.

(A_2) There exists $x^* \in D$ such that $F(x^*) = 0$ and $F'(x^*)^{-1} \in L(Y,X)$.

(A_3) There exists a function $\overline{W} : [0,\infty) \to [0,\infty)$ continuous and nondecreasing such that for each $x \in D$.

$$\|F'(x^*)^{-1}F''(x)\| \leq \overline{W}(\|x - x^*\|)$$

and $\overline{W}(0) = 0$.

(A_4) $\overline{U}(x^*, r) \subseteq D$,
where the radius of convergence r is the smallest positive zero of function $\Psi(t) = 0$ in the interval $(0, r_p)$, provided that equation

$$\overline{W}(t) = 1$$

has a minimal positive solution r_p for $\delta \neq 0$ or $W(0) \neq 0$,

$$\Psi(t) = \varphi(t) - 1,$$

and

$$\varphi(t) = \frac{\int_0^1 \overline{W}\big((1-\theta)t\big)(1-\theta)d\theta}{1-\overline{W}(t)}t.$$

Notice that $\Psi(0) = -1 < 0$, and $\Psi(t) \to \infty$ as $t \to r_p^-$. It follows by the intermediate value theorem that function Ψ has zeros in the interval $(0, r_p)$. We denote by r the smallest such zero.

Theorem 3.1 Suppose that the hypotheses (A) are satisfied. Then ,sequence $\{x_n\}$ generated for $x_0 \in U(x^*, r) - \{x^*\}$ by Newton's method is well defined in $U(x^*, r)$, remains in $U(x^*, r)$, and converges to x^* which is the only solution of equation $F(x) = 0$ in $\overline{U}(x^*, r)$. Moreover, the following estimates hold

$$\begin{aligned} \|x_{n+1} - x^*\| &\leq \frac{\int_0^1 \overline{W}\big((1-\theta)\|x_n - x^*\|\big)(1-\theta)d\theta}{1-\overline{W}(\|x_n - x^*\|)}\|x_n - x^*\|^2 \\ &\leq \|x_n - x^*\| < r. \end{aligned} \tag{3.1}$$

Proof. Let $x \in U(x^*, r)$. Using (A_2) and (A_3), we get in turn that

$$\begin{aligned} \|F'(x^*)^{-1}\Big(F'(x) - F'(x^*)\Big)\| &= \|F'(x^*)^{-1}\int_{x^*}^{x} F''(z)dz\| \\ &\leq \|\int_0^1 F'(x^*)^{-1}F''\Big(x^* + \theta(x-x^*)\Big)(x-x^*)d\theta\| \\ &\leq \int_0^1 \overline{W}(\theta\|x-x^*\|)d\theta\|x-x^*\| \\ &= \overline{W}(\|x-x^*\|) \leq \overline{W}(r) < 1. \end{aligned} \tag{3.2}$$

It follows from (3.1) and the Banach Lemma on invertible operators [18] that $F'(x)^{-1} \in L(Y, X)$, and

$$\|F'(x)^{-1}F'(x^*)\| \leq \frac{1}{1-\overline{W}(\|x-x^*\|)}. \tag{3.3}$$

In particular for $x = x_0$, point x_1 is well defined by Newton's method for $n = 0$. We can write by Newton's method for $n = 0$, and (A_2) that

$$\begin{aligned} x^* - x_1 &= x^* - x_0 + F'(x_0)^{-1}F(x_0) \\ &= -F'(x_0)^{-1}\Big(F(x^*) - F(x_0) - F'(x_0)(x^* - x_0)\Big) \\ &= -F'(x_0)^{-1}\int_{x_0}^{x^{**}} F''(z)(x^* - z)dz \\ &= -F'(x_0)^{-1}\Big(\int_0^1 F''\big(x_0 + \theta(x^* - x_0)\big)\Big)(x^* - x_0)^2(1-\theta)d\theta. \end{aligned} \tag{3.4}$$

Using $(A_2) - (A_4)$ and (3.4), we obtain in turn that

$$\begin{aligned} \|x_1 - x^*\| &\leq \|F'(x_0)^{-1}F'(x^*)\|\|\int_0^1 F'(x^*)^{-1}F''\big(x_0 + \theta(x^* - x_0)\big)(x_0 - x^*)^2(1-\theta)d\theta \\ &\leq \frac{\int_0^1 \overline{W}((1-\theta)\|x_0 - x^*\|)(1-\theta)d\theta}{1 - \overline{W}(\|x_0 - x^*\|)}\|x_0 - x^*\|^2 \\ &\leq \|x_0 - x^*\| < r, \end{aligned} \tag{3.5}$$

which shows (3.1) for $n = 0$ and $x_1 \in U(x^*, r)$. The induction for (3.1) is completed, if we replace x_0, x_1 by x_k, x_{k+1}, respectively in the preceding estimates. Then, from

$$\|x_{k+1} - x^*\| \leq q\|x_k - x^*\| < r,$$

where,

$$q = \frac{\int_0^1 \overline{W}\Big((1-\theta)\|x_0 - x^*\|\Big)(1-\theta)d\theta}{1 - \overline{W}(\|x_0 - x^*\|)} \in [0, 1),$$

we deduce that $\lim_{k\to\infty} x_k = x^*$, and $x_{k+1} \in U(x^*, r)$.

Finally, to show the uniqueness part, let $y^* \in \overline{U}(x^*, r)$ with $F(y^*) = 0$.

As in (3.5) with x_0, x^* replaced by x_k, y^*, we arrive at

$$\|x_{k+1} - y^*\| \leq \frac{\int_0^1 \overline{W}((1-\theta)\|x_k - y^*\|)(1-\theta)d\theta}{1 - \overline{W}(\|x_k - y^*\|)} \|x_k - y^*\|^2 \\ \leq q^* \|x_k - y^*\|,$$

where

$$q^* = \frac{\int_0^1 \overline{W}((1-\theta)\|x_0 - y^*\|)(1-\theta)d\theta}{1 - \overline{W}(\|x_0 - y^*\|)} \in [0,1).$$

Hence, we get that

$$\lim_{k\to\infty} x_k = y^*$$

But, we know that

$$\lim_{k\to\infty} x_k = x^*$$

Therefore, we conclude that $x^* = y^*$.

□

References

[1] Amat, S., Busquier, S. and Gutierrez, J.M., Geometric constructions of iterative functions to solve nonlinear equations, *J. Comput. Appl. Math.* 157, 2003, 197-205.

[2] Argyros, I.K., *Convergence and Applications of Newton-type Iterations*, Springer Verlag, Berlin, Germany, 2008.

[3] Argyros, I.K. and Magreñán, A.A., *A Contemporary Study of Iterative Methods*, Academic Press, Elsevier, San Diego, CA, USA, 2017.

[4] Denis, J.E. and Schnabel, R.B., *Numerical methods for unconstrained optimization and nonlinear equation*, SIAM, Philadelphia, 1996.

[5] Deulfhard, P. and Heindl, G., Affine invariant convergence theorems for Newton's method and extensions to related methods, *SIAM J. Numer. Anal.*, 16 (1979) 1 - 10.

[6] Ezquerro, J. A. and Hernández, M.A., A semi-local convergence result for Newton's method under generalized conditions of Kantorovich, *J. Complexity*, 30 (2014) 309 - 324.

[7] Ezquerro, J. A. and Hernández, M.A., Center conditions on high order derivates in the semi-local convergence of Newton's method, *J. Complexity*, 31 (2015) 277 - 292.

[8] Ezquerro, J. A. and Hernández M.A., Generalized differentiability conditions for Newton's method, *IMA J. Numer. Anal.*, 22, 2 (2002) 187 - 205.

[9] Ezquerro, J.A. and Hernández, M.A., Newton's method: an updated Approach of Kantorovich's theory, *Frontiers in Mathematics*, Birkhäuser, Cham Switzerland, (2017).

[10] Ezquerro, J. A. and Hernández M.A., On an application of Newton's method to non-linear operators with w-conditioned second derivative, *BIT*, 42, 3 (2002) 519 - 530.

[11] Ezquerro, J. A., González, D. and Hernández, M.A., Majorizing sequences for Newton's method from initial value problem, *J. Comput. Appl. Math.*, 236 (2012) 2246 - 2258.

[12] Gutiérrez, J. M., A new semi-local convergence theorem for Newton's method, *J. Comput. Appt. Math*, 79 (1997) 131-145.

[13] Gutiérrez, J. M. and Hernández, M.A., Majorizing sequences for Newton's method, *Tamkang J. Math.*, 29, 3 (1998) 199-202.

[14] Gutiérrez, J. M. and Hernández M.A., Newton's method under different Lipschitz conditions, (Numerical analysis and its applications) *Lectures Notes in Comput. Sci.*, 1988 (2000) 368 - 376.

[15] Gutiérrez, J. M. and Hernández, M.A., Newton's method under weak Kantorovich conditions, *IMA J. Numer. Anal.*, 20 (2000) 521 - 532.

[16] Hernández, M.A., Relaxing convergence conditions for Newton's method, *J. Math. Anal. Appl.*, 249 (2000) 463 - 475.

[17] Huang, Z., A note on the Kantorovich theorem for Newton method, *J. Comput. Appl. Math.*, 47 (1993) 211 - 217.

[18] Kantorovich, L.V. and Akilov, G. P., *Functional Analysis*, Pergamon Press, New York, 1982.

[19] Ortega, J. M. and Rheinboldt, W. C., *Iterative solutions of nonlinear equations in several variables*, Academic Press, New York, 1970.

Chapter 10

Convergence under Lipschitz Conditions: Case V

1. Introduction

The applicability of Newton's method given in [9, p. 128] for Banach space valued operators, and Lipschitz Fréchet derivative is extended without additional hypotheses.

2. Semi-Local Convergence Analysis

We first state the semi-local convergence analysis of Newton's method given in [9, p. 128] so we can compare it to the corresponding one given in Theorem 2.2 (see Remark 2.3).

The semi-local convergence of Newton's method given in [9] is based on the hypotheses (C):

(C_1) $F : D \subset X \to Y$ is continuously Fréchet-differentiable operator.
There exists a scalar function $\overline{g} \in C^2([0,\infty))$ such that $\overline{g}''(t) \geq 0$ for each $t \in [0,\infty)$ and satisfying the hypotheses:

(C_2) For some $x_0 \in D$

$$F'(x_0)^{-1} \in L(Y,X), \|F'(x_0)^{-1}F(x_0)\| \leq -\frac{\overline{g}(t)}{\overline{g}'(t)}$$

and

$$\|F'(x_0)^{-1}F^{(i)}(x_0)\| \leq b_i.$$

(C_3) $\|F'(x_0)^{-1}\Big(F'(x)-F'(y)\Big)\| \leq \overline{g}'(u_1)-\overline{g}'(u_2)$ for $\|x-y\| \leq u_2-u_1, x, y, \in D$ and $u_1, u_2 \in [0,\infty)$ with $u_1 \leq u_2$.

(C_4) There exists $\alpha \in (0,\infty)$ with $\overline{g}'(\alpha)=0, \overline{g}(t)$ *in* $(0,\alpha)$, and $\overline{g}(\alpha)=0$.

(C_5) $U(x_0,\overline{t}^*) \leq D$, where $\overline{t}^*$ is the smallest solution of equation $\overline{g}'(t)=0$ in $[0,\infty)$.

Theorem 2.1 [9] Suppose that the hypotheses (C) are satisfied. Then, the sequence $\{x_n\}$ generated for $x_0 \in D$ by Newton's method, is well defined in $\overline{U}(x_0,\overline{t}^*)$, remains in $\overline{U}(x_0,\overline{t}^*)$ for each $n=0,1,2,\ldots$, and converges to a solution $x^* \in \overline{U}(x_0,\overline{t}^*)$ of equation $F(x)=0$. Moreover, the following estimates hold

$$\|x_{n+1}-x_n\| \leq u_{n+1}-u_n$$

and

$$\|x_n-x^*\| \leq \overline{t}^*-u_n,$$

where

$$u_n = u_{n-1} - \frac{\overline{g}(u_{n-1})}{\overline{g}'(u_{n-1})}.$$

Furthermore, if the scalar function $\overline{g}(t)$ has two real zeros $\overline{t}^*$ and $\overline{t}^{**}$ such that $t_0 < \overline{t}^* \leq \overline{t}^{**}$, then the solution t^* is unique in $U(x_0,\overline{t}^{**}) \cap D$, if $\overline{t}^* < \overline{t}^{**}$ or in $\overline{U}(x_0,\overline{t}^*)$ if $\overline{t}^{**}=\overline{t}^*$.

The new semi-local convergence result is based on the hypotheses (H):

(H_1) = (C_1).

(H_2) There exist $x_0 \in D$ and $V:[0,\infty) \rightarrow [0,\infty)$ continuous, nondecreasing with $V(0)=0$ such that $F'(x_0)^{-1} \in L(Y,X)$ and for each

$$\|F'(x_0)^{-1}\Big(F'(x)-F'(x_0)\Big)\| \leq V(\|x-x_0\|)$$

Assume equation

$$V(t) = 1$$

has a minimal positive solution r_0.

Set

$$D_0 = D \cap U(x_0, r_0).$$

There exits a scalar function $g \in C^2([0,\infty)$ such that $g''(t) > 0$, and for each $t \in [0,\infty)$ satisfying the hypotheses:

(H_3)

$$\|F'(x_0)^{-1}F(x_0)\| \leq g(0)$$

and

$$\|F'(x_0)^{-1}\Big(F'(x) - F'(y)\Big)\| \leq g'(u_1) - g'(u_2)$$

for $\|x-y\| \leq u_2 - u_1, x, y \in D_0$ and $u_1, u_2 \in [0,\infty)$ with $u_1 \leq u_2$.

(H_4) There exists $\gamma \in (0,\infty)$ with $g'(\gamma) = 0$ and $g(\gamma) \leq 0$.

(H_5) $U(x_0, t^*) \subseteq D$, where t^* is the smallest solution of equation $g(t) = 0$ in $(0,\gamma)$.

(H_6) $V(t) \leq g'(t) + 1$ for each $t \in [0, t^*)$.

(H_7) There exists $\gamma_1 \geq \gamma$ such that

$$\int_0^1 V\Big((1-\theta)\gamma + \theta\gamma_1\Big)d\theta < 1.$$

Set $D_1 = D \cap \overline{U}(x_0, \gamma_1)$.

Theorem 2.2 Suppose that the hypotheses (H) are satisfied. Then, sequence $\{x_n\}$ generated for $x_0 \in D$ by Newton's method is well defined in $\overline{U}(x_0, t^*)$, remains in $U(x_0, t^* - t_0)$ for each $n = 0, 1, 2, \ldots$, and converges to a solution $x^* \in \overline{U}(x_0, t^* - t_0)$ of equation $F(x) = 0$.

Moreover, the following estimates hold for each $n = 0, 1, 2, \ldots$

$$\|x_{n+1} - x_n\| \leq t_{n+1} - t_n$$

and

$$\|x_n - x^*\| \leq t^* - t_n,$$

where

$$t_n = t_{n-1} - \frac{g(t_{n-1})}{g'(t_{n-1})}.$$

Furthermore, the point x^* is the only solution of equation $F(x) = 0$ in D_1.

Proof. Let $x \in U(x_0, r_0)$. Using (H_2), we have that

$$\|F'(x_0)^{-1}\Big(F'(x) - F'(x_0)\Big)\| \leq V(\|x - x_0\|) \leq V(r_0) < 1. \tag{2.1}$$

Then, it follows from (2.1) and the Banach lemma on invertible operators [18] that $F'(x)^{-1} \in L(Y, X)$, and

$$\|F'(x)^{-1}F'(x_0)\| \leq \frac{1}{1 - V(\|x - x_0\|)}. \tag{2.2}$$

By (H_6) and (2.2) we get that

$$\|F'(x)^{-1}F'(x_0)\| \leq \frac{1}{1 - V(\|x - x_0\|)} \leq -\frac{1}{g'(\|x - x_0\|)}. \tag{2.3}$$

The rest of the proof until the uniqueness part follows exactly as in the proof given in [9], and by simply noticing that the iterates $\{x_n\}$ remains in D_0 which is at least as precise domain as D, since $D_0 \subseteq D$. To show the uniqueness part, let $y^* \in D_1$ with $F(y^*) = 0$. Define operator M by

$$M = \int_0^1 F'\Big(x^* + \theta(y^* - x^*)\Big)d\theta.$$

Using (H_2) and (H_7), we obtain in turn that

$$\|F'(x_0)^{-1}\Big(M - F'(x_0)\Big)\| \leq \int_0^1 v\Big(\|x^* + \theta(y^* - x^*) - x_0\|\Big)d\theta$$
$$\leq \int_0^1 v\Big((1-\theta)\|x^* - x_0\| + \theta\|y^* - x_0\|\Big)$$
$$\leq \int_0^1 v\Big((1-\theta)\gamma + \theta\gamma_1\Big)d\theta < 1,$$

so $M^{-1} \in L(Y,X)$. Finally, using the identity

$$0 = F(y^*) - F(x^*) = M(y^* - x^*),$$

we deduce that $x^* = y^*$.

□

Remark 2.3 (i) Suppose that hypotheses of Theorem 2.1 and Theorem 2.2 hold. Then, we have for each $t \in [t_0, \min\{\alpha, \gamma\})$

$$g'(t) \leq \overline{g}'(t). \tag{2.4}$$

If for each $t_0 < s \leq t$

$$-\frac{g(s)}{g'(s)} \leq -\frac{\overline{g}(t)}{\overline{g}'(t)}$$

and

$$s - \frac{g(s)}{g'(s)} \leq t - \frac{\overline{g}(t)}{\overline{g}'(t)},$$

then, a simple inductive argument shows that

$$t_n \leq u_n,$$

$$t_{n+1} - t_n \leq u_{n+1} - u_n,$$

so,

$$t^* \leq \overline{t}^*.$$

Hence, the error bounds on the distances are at least as tight and the information on the location of the solution x^* is at least as precise.

(ii) Clearly, in view of (2.4), the hypotheses of Theorem 2.1 imply the hypotheses of Theorem 1.2 but not necessarily vice versa.
For example from (2.4)

$$g(\alpha) \leq \overline{g}(\alpha) \leq 0$$

the existence of γ is implied.

(iii) The results obtained here can be improved even further, if we replace the set D_0 by the set $\tilde{D_0}$ defined by

$$\tilde{D_0} = D \cap U(x_1, r_0 - \eta)$$

provided that $\eta < r_0$.

Notice that the iterated $\{x_n\}$ lie in $\tilde{D_0}$ which is a more precise location that D_0, since $\tilde{D_0} \subseteq D_0$. The new function $\tilde{W}$ replacing W in hypotheses (H) and Theorem 2.2 will be at least as tight as W leading to the least as weak sufficient semi-local convergence conditions for Newton's method. It is worth noticing that the advantages $(i)-(iii)$ are obtained under the same computational cost as in Theorem 2.1, since the computation of Lipschitz-type condition (C_3) passes through the computation of conditions (H_3) and condition (H_4) as special cases.

Remark 2.4 Functions $\overline{g}, g$ and $V(t)$ can be chosen by

$$\overline{g}(t) = \frac{L}{2}t^2 - t + \eta,$$

$$g(t) = \frac{K}{2}t^2 - t + \eta,$$

and

$$V(t) = L_0 t,$$

where

$$\|F'(x_0)^{-1}F(x_0)\| \leq \eta,$$

$$\|F'(x_0)\Big(F'(x) - F'(y)\Big)\| \leq L\|x - y\| \text{ for each } x, y \in D$$

$$\|F'(x_0)\Big(F'(x)-F'(x_0)\Big)\| \le L_0\|x-x_0\| \text{ for each } x\in D$$

and

$$\|F'(x_0)\Big(F'(x)-F'(y)\Big)\| \le K\|x-y\| \text{ for each } x,y\in D_0 = D\cap U(x_0,\frac{1}{L_0})$$

Then, we have

$$K \le L \text{ and } L_0 \le L.$$

Moreover, the sufficient convergence criteria for $L_0 \le K$ corresponding to Theorem 2.1 and Theorem 2.2 are

$$H_K = L\eta \le \frac{1}{2} \tag{2.5}$$

respectively. Then, we have that

$$H_K \le \frac{1}{2} \Rightarrow H_A \le \frac{1}{2} \tag{2.6}$$

but not necessarily vice versa, unless if $K = L$.
Hypothesis (2.5) is due to Kantorovich. In view of (2.5) this hypothesis is improved, if $K < L$ (such examples can be found in [2,3]).

3. Local Convergence Analysis

We first base the local convergence analysis (not given in [9]) on the hypotheses (A):

(A_1) $F : D \subset X \to Y$ is a continuously Fréchet-differentiable operator.

(A_2) There exist $x^* \in D$ such that

$$F(x^*) = 0 \text{ and } F'(x^*)^{-1} \in L(Y,X)$$

(A_3) There exists a function $V : [0,\infty) \to [0,\infty)$ continuous and nondecreasing such that for each $x \in D$

$$\|F'(x^*)^{-1}\Big(F(x)-F(x^*)\Big)\| \le V(\|x-x^*\|)$$

and $V(0) = 0$.

(A_4) Equation $V(t)=1$ has a minimal positive solution r_0.
Set

$$D_0 = D \cap U(x^*, r_0)$$

(A_5)

$$\|F'(x^*)^{-1}\Big(F'(y)-F'(x)\Big)\| \leq W(\|y-x\|)$$

for each $x, y \in D_0$ and $W : [0, r_0) \to [0, \infty)$ be continuous, and nondecreasing function with $W(0)=0$.
Define functions φ and Ψ on $(0, r_0)$ by

$$\varphi(t) = \frac{\int_0^1 W\Big((1-\theta)t\Big)d\theta}{1-V(t)},$$

and

$$\Psi(t) = \varphi(t) - 1.$$

Notice that $\Psi(0) = -1 < 0$, and $\Psi(t) \to \infty$ as $t \to r_0^-$. It follows by the intermediate value theorem that function Ψ has zeros in the interval $(0, r_0)$. We denote by r the smallest such zero.

Theorem 3.1 Suppose that the hypotheses (A) are satisfied. Then, sequence $\{x_n\}$ generated for $x_0 \in U(x^*, r) - \{x^*\}$ by Newton's method is well defined in $U(x^*, r)$, remains in $U(x^*, r)$, and converges to $x*$, which is the only solution of equation $F(x) = 0$ in $\overline{U}(x^*, r_0)$. Moreover, the following estimates hold

$$\|x_{n+1} - x^*\| \leq \frac{\int_0^1 W\Big((1-\theta)\|x_n - x^*\|\Big)d\theta}{1-V(\|x_n - x^*\|)}\|x_n - x^*\| \leq \|x_n - x^*\| < r. \quad (3.1)$$

Proof. Let $x \in U(x^*, r)$. Using (A_2) and (A_3), we get in turn that

$$\|F'(x^*)^{-1}\Big(F'(x) - F'(x^*)\Big)\| \leq V(\|x - x^*\|) \leq p(r) < 1. \quad (3.2)$$

It follows from (3.1) and the Banach lemma on invertible operators [18] that $F'(x)^{-1} \in L(Y, X)$, and

$$\|F'(x)^{-1}F'(x^*)\| \leq \frac{1}{1-V(\|x-x^*\|)}. \quad (3.3)$$

In particular for $x = x_0$, point x_1 is well defined by Newton's method for $n = 0$. We can write by Newton's method for $n = 0$ and (A_2) that

$$\begin{aligned} x^* - x_1 &= x^* - x_0 + F'(x_0)^{-1} F(x_0) \\ &= -F'(x_0)^{-1}\Big(F(x^*) - F(x_0) - F'(x_0)(x^* - x_0)\Big) \end{aligned} \tag{3.4}$$

Using $(A_2) - (A_4)$ and (3.4), we obtain in turn that

$$\begin{aligned} \|x_1 - x^*\| &\le \|F'(x_0)^{-1}F'(x^*)\| \|\int_0^1 F'(x^*)^{-1}\Big(F'(x^* + \theta(x^* - x_0)) - F(x_0)\Big)(x_0 - x^*)d\theta\| \\ &\le \frac{\int_0^1 W\Big((1-\theta)\|x_0 - x^*\|\Big)d\theta}{1 - V(\|x_0 - x^*\|)}\|x_0 - x^*\| \le \|x_0 - x^*\| < r, \end{aligned} \tag{3.5}$$

which shows (3.1) for $n = 0$ and $x_1 \in U(x^*, r)$. The induction for (3.1) is completed, if we related x_0, x_1 by x_k, x_{k+1} in the preceding estimates. Then, from

$$\|x_{k+1} - x^*\| \le q\|x_k - x^*\| < r,$$

where

$$q = \frac{\int_0^1 W\Big((1-\theta)\|x_0 - x^*\|\Big)d\theta}{1 - V(\|x_0 - x^*\|)} \in [0, 1),$$

we deduce that

$$\lim_{k \to \infty} x_k = x^*$$

and $x_{k+1} \in U(x^*, r)$.
The uniqueness part is standard [2,3]. □

References

[1] Amat, S., Busquier, S. and Gutierrez, J.M., Geometric constructions of iterative functions to solve nonlinear equations, *J. Comput. Appl. Math.* 157, 2003, 197-205.

[2] Argyros, I.K., *Convergence and Applications of Newton-type Iterations*, Springer Verlag, Berlin, Germany, 2008.

[3] Argyros, I.K. and Magreñán, A.A., *A Contemporary Study of Iterative Methods*, Academic Press, Elsevier, San Diego, CA, USA, 2017.

[4] Denis, J.E. and Schnabel, R.B., *Numerical methods for unconstrained optimization and nonlinear equation*, SIAM, Philadelphia, 1996.

[5] Deulfhard, P. and Heindl, G., Affine invariant convergence theorems for Newton's method and extensions to related methods, *SIAM J. Numer. Anal.*, 16 (1979) 1 - 10.

[6] Ezquerro, J. A. and Hernández, M.A., A semi-local convergence result for Newton's method under generalized conditions of Kantorovich, *J. Complexity*, 30 (2014) 309 - 324.

[7] Ezquerro, J. A. and Hernández, M.A., Center conditions on high order derivates in the semi-local convergence of Newton's method, *J. Complexity*, 31 (2015) 277 - 292.

[8] Ezquerro, J. A. and Hernández M.A., Generalized differentiability conditions for Newton's method, *IMA J. Numer. Anal.*, 22, 2 (2002) 187 - 205.

[9] Ezquerro, J.A. and Hernández, M.A., Newton's method: an updated Approach of Kantorovich's theory, *Frontiers in Mathematics*, Birkhäuser, Cham Switzerland, (2017).

[10] Ezquerro, J. A. and Hernández M.A., On an application of Newton's method to non-linear operators with w-conditioned second derivative, *BIT*, 42, 3 (2002) 519 - 530.

[11] Ezquerro, J. A., González, D. and Hernández, M.A., Majorizing sequences for Newton's method from initial value problem, *J. Comput. Appl. Math.*, 236 (2012) 2246 - 2258.

[12] Gutiérrez, J. M., A new semi-local convergence theorem for Newton's method, *J. Comput. Appt. Math*, 79 (1997) 131-145.

[13] Gutiérrez, J. M. and Hernández, M.A., Majorizing sequences for Newton's method, *Tamkang J. Math.*, 29, 3 (1998) 199-202.

[14] Gutiérrez, J. M. and Hernández M.A., Newton's method under different Lipschitz conditions, (Numerical analysis and its applications) *Lectures Notes in Comput. Sci.*, 1988 (2000) 368 - 376.

[15] Gutiérrez, J. M. and Hernández, M.A., Newton's method under weak Kantorovich conditions, *IMA J. Numer. Anal.*, 20 (2000) 521 - 532.

[16] Hernández, M.A., Relaxing convergence conditions for Newton's method, *J. Math. Anal. Appl.*, 249 (2000) 463 - 475.

[17] Huang, Z., A note on the Kantorovich theorem for Newton method, *J. Comput. Appl. Math.*, 47 (1993) 211 - 217.

[18] Kantorovich, L.V. and Akilov, G. P., *FunctionalAnalysis*, Pergamon Press, New York, 1982.

[19] Ortega, J. M. and Rheinboldt, W. C., *Iterative solutions of nonlinear equations in several variables*, Academic Press, New York, 1970.

Chapter 11

Convergence under w-Lipschitz Conditions: Case VI

1. Introduction

The applicability of Newton's method given in [9. ch.4, p. 133] is extended for Banach space valued operators, and w-Lipschitz conditions without additional conditions.

2. Semi-Local Convergence Analysis

The semi-local convergence analysis of Newton's method is based on the hypotheses (A):

(A_1) $F : D \subset X \to Y$ is a continuously Fréchet-differentiable operator.

(A_2) There exist $x_0 \in D, \eta \geq 0$ and $V : [0,\infty) \to [0,\infty)$ continuous and nondecreasing with $V(0) = 0$ such that for each $x \in D$

$$F'(x_0)^{-1} \in D, \|F'(x_0)^{-1}F(x_0)\| \leq \eta$$

and

$$\|F'(x_0)^{-1}\Big(F'(x) - F'(x_0)\Big)\| \leq V(\|x - x_0\|).$$

Assume equation $V(t) = 1$ has a minimal positive solution r_0.
Set

$$D_0 = D \cap U(x_0, r_0).$$

Suppose that $\eta \leq r_0$.

(A_3) There exists $W : [0, r_0) \to [0, \infty)$ continuous and nondecreasing with $W(0) = 0$ such that for each $x, y \in D_0$

$$\|F'(x_0)^{-1}\Big(F'(x) - F'(y)\Big)\| \leq W(\|x-y\|).$$

Define function $\varphi : [\eta, r_0) \to \mathbb{R}$ by

$$\varphi(t) = td(t) + (1 - d(t))(1 + d_0)\eta + d_0\eta - t,$$

where

$$d_0 = \frac{\int_0^1 W(\theta\eta)d\theta}{1 - V(\eta)},$$

and

$$d(t) = \frac{\int_0^1 W(\theta\eta)d\theta}{1 - V(t)}.$$

(A_4) Suppose equation $\varphi(t) = 0$ has positive zeros in the interval $(\eta, r_0]$. Denote by r the smallest such zero.

(A_5) $d(r) < 1$.

(A_6) $\overline{U}(x_0, r) \subseteq D$.

(A_7) $r_1 \geq r$ and

$$\int_0^1 V(\theta r_1 + (1-\theta)r)d\theta < 1.$$

Define the set $D_1 = D \cap U(x_0, r_1)$.

Theorem 2.1 Suppose that the hypotheses (A) are satisfied. Then, sequence $\{x_n\}$ generated by Newton's method starting at x_0 is well defined in $U(x_0,r)$, remains in $U(x_0,r)$ for each $n = 0,1,2,\ldots$, and converges to a unique solution x^* of equation $F(x) = 0$ in $\overline{U}(x_0,r)$.
Moreover, the following estimates hold

$$\|x_2 - x_1\| \leq d_0\|x_1 - x_0\|, \tag{2.1}$$

$$\|x_{n+1} - x_n\| \leq d(r)^{n-1}\|x_2 - x_1\|, n = 2,3,\ldots. \tag{2.2}$$

and

$$\|x_n - x^*\| \leq \frac{d^{n-1}(r)}{1-d(r)} d_0\eta, n = 3,4,\ldots. \tag{2.3}$$

Furthermore, the limits point x^* is the only solution of equation $F(x) = 0$ in D_1.

Proof. Let $x \in U(x_0,r)$. Using (A_2), we have that

$$\|F'(x_0)^{-1}\Big(F'(x) - F'(x_0)\Big)\| \leq V(\|x - x_0\|) \leq V(r) < 1. \tag{2.4}$$

It follows from (2.4) and the Banach lemma on invertible operators [18] that $F'(x)^{-1} \in L(Y,X)$, and

$$\|F'(x)^{-1}F'(x_0)\| \leq \frac{1}{1 - V(\|x - x_0\|)}. \tag{2.5}$$

In particular for $x = x_1$ and $\|x_1 - x_0\| \leq \eta < r$, we get that x_2 is well defined by Newton's method for $n = 0$. We can write by Newton's method for $n = 0$.

$$\begin{aligned} F(x_1) &= F(x_1) - F(x_0) - F'(x_0)(x_1 - x_0) \\ &= \int_0^1 \Big(F'(x_0 + \theta(x_1 - x_0)) - F'(x_0)\Big)(x_1 - x_0)d\theta, \end{aligned} \tag{2.6}$$

so by (A_3)

$$\begin{aligned}\|F'(x_0)^{-1}F(x_1)\| &= \|\int_0^1 F'(x_0)^{-1}\Big(F'(x_0+\theta(x_1-x_0))-F'(x_0)\Big)(x_1-x_0)d\theta\| \\ &\leq \int_0^1 W(\theta\|x_1-x_0\|)d\theta\|x_1-x_0\| \\ &\leq \int_0^1 W(\theta\eta)d\theta\eta. \end{aligned} \tag{2.7}$$

In view of (2.2) (for $x=x_1$), (2.4) and Newton's method for $n=1$, we obtain in turn that

$$\begin{aligned}\|x_2-x_1\| &= \|F'(x_1)^{-1}F'(x_0)F'(x_0)^{-1}F(x_1)\| \\ &\leq \frac{\int_0^1 W(\theta\|x_1-x_0\|)d\theta\|x_1-x_0\|}{1-V(\|x_1-x_0\|)} \\ &= d_0\|x_1-x_0\|\end{aligned}$$

and

$$\|x_2-x_0\| \leq \|x_2-x_1\|+\|x_1-x_0\| \leq (1+d_0)\|x_1-x_0\| \leq (1+d_0)\eta < r, \tag{2.8}$$

by the definition of r, so (2.1), $x_2 \in U(x_0,r)$ hold and x_3 is well defined by (2.5) (for $x=x_2$).
As in (2.6) we can write for $k=1,2,3,\ldots$

$$\begin{aligned}F(x_{k+1}) &= F(x_{k+1})-F(x_k)-F'(x_k)(x_{k+1}-x_k) \\ &= \int_0^1 \Big(F'(x_k+\theta(x_{k+1}-x_k))-F'(x_k)\Big)(x_{k+1}-x_k)d\theta.\end{aligned}$$

so

$$\|F'(x_0)^{-1}F(x_{k+1})\| \leq \int_0^1 W(\theta\|x_{k+1}-x_k\|)d\theta\|x_{k+1}-x_k\|$$
$$\leq \int_0^1 W(\theta\eta)d\theta\|x_{k+1}-x_k\|. \tag{2.9}$$

Then, by (2.5) (for $x = x_{n+1}$) and (2.9)

$$\begin{aligned}\|x_{k+2}-x_{k+1}\| &\leq \|F'(x_{k+1})^{-1}F'(x_0))\|\|F'(x_0)^{-1}F(x_{k+1})\| \\ &\leq \frac{\int_0^1 W(\theta\eta)d\theta\|x_{k+1}-x_k\|}{1-V(\|x_{k+1}-x_0\|)} \\ &\leq d(r)\|x_{k+1}-x_k\| \leq d(r)^{k-1}\|x_2-x_1\|\end{aligned} \tag{2.10}$$

which shows (2.2) and

$$\begin{aligned}\|x_{k+2}-x_0\| &\leq \|x_{k+2}-x_{k+1}\|+\cdots+\|x_1-x_0\| \\ &\leq \frac{1-d^{k-1}(r)}{1-d(r)}\|x_2-x_1\|+(1+d_0)\eta \\ &\leq \frac{1}{1-d(r)}d_0\eta+(1+d_0)\eta = r,\end{aligned} \tag{2.11}$$

so $x_{k+2} \in U(x_0, r)$.
We also have the estimate for $m = 1, 2, \ldots, k = 2, 3, \ldots$.

$$\begin{aligned}\|x_{k+m}-x_k\| &\leq \|x_{k+m}-x_{k+m-1}\|+\|x_{k+m-1}-x_{k+m-2}\|+\cdots+\|x_{k+1}-x_k\| \\ &\leq \left(d(r)^{k+m-2}+\cdots+d(r)^{k-1}\right)\|x_2-x_1\| \\ &(1+\cdots+d(r)^{m-1})d(r)^{k-1}\|x_2-x_1\| \\ &= \frac{1-d(r)^m}{1-d(r)}d^{k-1}(r)d_0\eta.\end{aligned} \tag{2.12}$$

By (2.12) sequence $\{x_k\}$ is Cauchy in a Banach space X and as such it converges to some $x^* \in \overline{U}(x_0, r)$ (since $\overline{U}(x_0, r)$ is a closed set). By letting $k \to \infty$ in (2.9), we deduce that $F(x^*) = 0$. Estimate (2.3) is obtained by letting $m \to \infty$ in (2.12).

Finally, to show the uniqueness part, let $y^* \in D_1$ with $F(y^*) = 0$.
Define

$$T = \int_0^1 F'(x^* + \theta(y^* - x^*))d\theta.$$

Then, we get by (A_2) and (A_6)

$$\begin{aligned} \|F'(x_0)^{-1}(T - F'(x_0))\| &\leq \int_0^1 V\Big(\theta\|y^* - x_0\| + (1-\theta)\|x^* - x_0\|\Big)d\theta \\ &\leq \int_0^1 V(\theta r_1 + (1-\theta)r)d\theta < 1, \end{aligned} \tag{2.13}$$

so $T^{-1} \in L(Y, X)$. Then. from the identity $0 = F(y^*) - F(x^*) = T(y^* - x^*)$, we conclude that $x^* = y^*$

□

Remark 2.2 **(a)** Hypothesis (A_2) is used to define r_0 and D_0, which may be a strict subset of D. Then, function W is defined using D_0 not D. This way we obtain a more precise location D_0 where the iterates $\{x_n\}$ lie than D. In the literature hypothesis
(A_3)' There exists $\overline{W} : [0, \infty) \to [0, \infty)$ continuous and nondecreasing with $\overline{W}(0) = 0$ such that for each $x, y \in D$

$$\|F'(x_0)^{-1}\Big(F'(x) - F'(y)\Big)\| \leq \overline{W}(\|x - y\|),$$

instead of (A_3). Moreover, (A_2) is not used and $\overline{W}(t) = V(t)$.
This way

$$W(t) \leq \overline{W}(t)$$

and

$$V(t) \leq \overline{W}(t)$$

leading to weaker sufficient convergence conditions, more precise error bounds on the distances $\|x_{n+1} - x_n\|, \|x_n - x^*\|$ and an at least as precise information on the location of the solution.

(b) The results can be extended if D_0 is replaced by $\tilde{D_0} \subseteq D_0$, where

$$\tilde{D_0} = D \cap U(x_1, r_0 - \eta), r_0 \geq \eta,$$

since the resulting function W will be even tighter than the function W depending on D_0. Notice that in this case we are still using the initial data (x_0, F, F'). Numerical examples where the inequalities are strict can be found in [2,3].

3. Local Convergence Analysis

The local convergence analysis of Newton's method is baed on hypotheses (A):

(A_1) $F : D \subset X \to Y$ is a nonlinear $k(k \geq 3)$ times continuously Fréchet-differentiable operator.

(A_2) There exist $x^* \in D$ and $b_i \geq 0, i = 3, 4, \ldots, k$ such that $F(x^*) = 0, F'(x^*)^{-1} \in L(Y, X)$ and

$$\|F'(x^*)^{-1} F^{(i)}(x^*)\| \leq b_i.$$

(A_3) There exists a function $V : [0, \infty) \to [0, \infty)$ continuous and nondecreasing with $V(0) = 0$ such that for each $x \in D$

$$\|F'(x^*)^{-1}\Big(F'(x) - F'(x^*)\Big)\| \leq V(\|x - x^*\|).$$

Assume equation $V(t) = 1$ has a minimal positive solution r_0.
Set

$$D_0 = D \cap U(x^*, r_0).$$

(A_4) There exists a function $W : [0, r_0) \to [0, \infty)$ continuous and nondecreasing with $W(0) = 0$ such that for each $x \in D_0$.

$$\|F'(x^*)^{-1}\left(F^{(k)}(x) - F^{(k)}(x^*)\right)\| \le W(\|x - x^*\|).$$

Define functions h and h_0 on the interval $[0, r_0)$ by

$$h(t) = \frac{\sum\limits_{i=2}^{k} \frac{1}{(i-1)!} b_i t^{i-1} + \frac{1}{(k-3)!} \int\limits_0^1 \int\limits_0^1 W(\theta(1-\tau)t)(1-\theta)^{k-3}(1-\tau)t^{k-1} d\theta d\tau}{1 - V(t)}$$

and

$$h_0(t) = h(t) - 1.$$

Notice that $h_0(0) = -1 < 0$ and $h_0(t) \to \infty$ as $t \to r_0^-$. Then, it follows from the intermediate value theorem that function h_0 has zeros in the interval $(0, r_0)$. Denote by r the smallest such zero.

(A_5) $\overline{U}(x^*, r) \subseteq D$.

(A_6) There exists $r^* \ge r$ such that

$$\int\limits_0^1 V(\theta r^*) d\theta < 1.$$

Define the set $D_1 = D \cap \overline{U}(x^*, r^*)$.

Next, we can show the following local convergence analysis of Newton's method using the preceding notation and the hypotheses (A).

Theorem 3.1 Suppose that the hypotheses (A) are satisfied. Then, sequence $\{x_n\}$ generated for $x_0 \in U(x^*, r) - \{x^*\}$ by Newton's method is well defined in $U(x^*, r)$ and remains in $U(x^*, r)$ for each $n = 0, 1, \ldots$, and converges to x^*, which is the only solution of equation $F(x) = 0$ in D_1. Moreover, the following estimates hold

$$\|x_{n+1} - x^*\| \le h(\|x_n - x^*\|)\|x_n - x^*\| \le \|x_n - x^*\| < r. \tag{3.1}$$

Proof. We shall use mathematical induction on m to show estimate (3.1). Let $x \in U(x^*, r_0)$ and $k = 3, 4, \ldots$. By Taylor series about x^*, we obtain the following identity

$$\begin{aligned} F''(x) &= \sum_{i=2}^{k-1} \frac{1}{(i-2)!} F^{(i)}(x^*)(x-x^*)^{i-2} + \frac{1}{(k-3)!} \int_{x^*}^{x} F^{(k)}(z)(x-z)^{k-3} dz \\ &= \sum_{i=2}^{k-1} \frac{1}{(i-2)!} F^{(i)}(x^*)(x-x^*)^{i-2} + \frac{1}{(k-3)!} \int_{x^*}^{x} \left(F^{(k)}(z) \pm F^{(k)}(x^*)\right)(x-z)^{k-3} dz \\ &= \sum_{i=2}^{k-1} \frac{1}{(i-2)!} F^{(i)}(x^*)(x-x^*)^{i-2} \\ &+ \frac{1}{(k-3)!} \int_{0}^{1} \left(F^{(k)}(x^* + \theta(x-x^*)) - F^{(k)}(x^*)\right)(x-x^*)^{k-2}(1-\theta)^{k-3} d\theta. \end{aligned} \tag{3.2}$$

Using the definition of r_0 and hypotheses (A_3) we get that

$$\|F'(x^*)^{-1}\left(F'(x) - F'(x^*)\right)\| \leq V(\|x-x^*\|) \leq V(r) \leq V(r_0) < 1. \tag{3.3}$$

It follows from (3.3), and the Banach lemma on invertible operators [18] that

$$F'(x)^{-1} \in L(Y, X)$$

and

$$\|F'(x)^{-1} F'(x^*)\| \leq \frac{1}{1 - V(\|x-x^*\|)}. \tag{3.4}$$

In particular for $x = x_0, x_1$ is well defined by Newton's method for $n = 0$. Suppose x_m are well defined for $m = 0, 1, 2, \ldots, n$.

In view of Newton's method, we have the approximation

$$\begin{aligned} x^* - x_{m+1} &= x^* - x_m + F'(x_m)^{-1}F(x_m) \\ &= -F'(x_m)^{-1}\Big[F(x^*) - F(x_m) - F'(x_m)(x^* - x_m)\Big] \\ &= -F'(x_m)^{-1}\int_{x_m}^{x^*} F''(z)(x^* - z)dz \\ &= -F'(x_m)^{-1}F'(x^*)\int_0^1 F'(x^*)^{-1}F''\Big(x_m + \tau(x^* - x_m)\Big)\times \\ &\quad (x^* - x_m)^2(1-\tau)dz, \end{aligned} \tag{3.5}$$

so

$$\|x_{m+1} - x^*\| \leq \|F'(x_m)^{-1}F'(x^*)\| \Big\|\int_0^1 F'(x^*)^{-1}F''\Big(x_m + \tau(x^* - x_m)\Big)\times (x^* - x_m)^2(1-\tau)dz\Big\| \tag{3.6}$$

We need an upper bound on $\|F'(x^*)^{-1}F''(x)\|$. To find such a bound we use hypotheses $(A_2), (A_4)$ and (3.2) to obtain in turn that

$$\begin{aligned} \|F'(x^*)^{-1}F''(x)\| &\leq \sum_{i=2}^{k} \frac{1}{(i-2)!}\|F'(x^*)^{-1}F^{(i)}(x^*)\|\|x - x^*\|^{i-2} \\ &\quad + \frac{1}{(k-3)!}\int_0^1 \|F'(x^*)^{-1}\Big(F^{(k)}(x^* + \theta(x - x^*)) - F^{(k)}(x^*)\Big)\|\times \\ &\quad \|x - x^*\|^{k-2}(1-\theta)^{k-3}d\theta \\ &\leq \sum_{i=2}^{k} \frac{1}{(i-2)!}\|F'(x^*)^{-1}F^{(i)}(x^*)\|\|x - x^*\|^{i-2} \\ &\quad + \frac{1}{(k-3)!}\int_0^1 W(\theta\|x - x^*\|)\|x - x^*\|^{k-2}(1-\theta)^{k-3}d\theta. \end{aligned} \tag{3.7}$$

In particular, for $x = x_m + \tau(x^* - x_m)$, we obtain from (3.7) that

$$\begin{aligned}
&\|F'(x^*)^{-1}F''(x_m + \tau(x^* - x_m))\| \\
&\leq \sum_{i=2}^{k} \frac{1}{(i-2)!} b_i \int_0^1 (1-\tau)^{i-2} \|x^* - x_m\|^{i-2} \|x^* - x_m\|^2 d\tau \\
&+ \frac{1}{(k-3)!} \int_0^1 \int_0^1 w\Big(\theta(1-\tau)\|x^* - x_m\|\Big)(1-\tau)^{k-2}\|x^* - x_m\|^{k-2} \times \\
&\|x^* - x_m\|^2 (1-\theta)^{k-3} d\theta d\tau \\
&\leq \sum_{i=2}^{k} \frac{1}{(i-1)!} b_i \|x^* - x_m\|^i \\
&+ \frac{1}{(k-3)!} \int_0^1 \int_0^1 w\Big(\theta(1-\tau)\|x^* - x_m\|\Big)(1-\tau)^{k-2}(1-\theta)^{k-3} d\theta d\tau \|x^* - x_m\|^k. \qquad (3.8)
\end{aligned}$$

By combining (3.4) and (3.8) in (3.6) and using the definition of r and function h we get in turn that

$$\|x_{m+1} - x^*\| \leq h(\|x^* - x_m\|)\|x^* - x_m\| \leq \|x^* - x_m\| < r, \qquad (3.9)$$

which completes the induction for (3.1). Then, from the estimate

$$\|x_{m+1} - x^*\| \leq q\|x_m - x^*\| < r \qquad (3.10)$$

where

$$q = h(\|x^* - x_0\|) \in [0, 1).$$

We deduce that

$$\lim_{m \to \infty} x_m = x^* \text{ and } x_{m+1} \in U(x^*, r).$$

Finally, to show the uniqueness part, let $y^* \in D_1$ with $F(y^*) = 0$. Define the linear operator

$$T = \int_0^1 V\Big(\theta\|y^* - x^*\|\Big) d\theta.$$

Using (A_3) we get that

$$\|F'(x^*)^{-1}\Big(T - F'(x^*)\Big)\| \leq \int_0^1 v\Big(\theta\|y^* - x^*\|\Big) d\theta \leq \int_0^1 v(\theta r^*) d\theta < 1, \qquad (3.11)$$

so $T^{-1} \in L(Y,X)$. Then, from the approximation

$$0 = F(y^*) - F(x^*) = T(y^* - x^*), \tag{3.12}$$

we conclude that $x^* = y^*$.

□

The case $k = 2$ is studied next, using the hypotheses (B):

(B_1) $F : D \subset X \to Y$ is a twice continuously Fréchet-differentiable operator.

(B_2) There exists $x^* \in D$ and $b_2 \geq 0$ such that

$$F(x^*) = 0, F'(x^*)^{-1} \in L(Y,X)$$

and

$$\|F'(x^*)^{-1} F''(x^*)\| \leq b_2.$$

(B_3) = (A_3)

(B_4) = (B_4) for $k = 2$.

Define functions h and h_0 on the interval $[0, r_0)$ by

$$h(t) = \frac{\int_0^1 W\Big((1-\theta)t\Big) d\theta + \frac{1}{2} b_2}{1 - V(t)} t$$

and

$$h_0(t) = h(t) - 1.$$

Denote again by ρ the smallest zero of function h in the interval $(0, r_0)$.

(B_5) $\overline{U}(x^*, \rho) \subseteq D$.

(B_6) There exists $\rho^* \geq \rho$ such that

$$\int_0^1 V(\theta \rho^*) d\theta < 1.$$

Define the set $D_1 = D \cap U(x^*, \rho^*)$.

Using the preceding notation we arrive at:

Theorem 3.2 Suppose that the hypotheses (B) are satisfied. Then, the conclusion of Theorem 3.1 hold.

Proof. Notice that we obtain the numerator at the definition of function h as follows:

$$\begin{aligned}
&\|\int_0^1 F'(x^*)^{-1}\Big(F''(x_m+\tau(x^*-x_m))\Big)(1-\tau)(x^*-x_m)^2 d\tau\| \\
&\leq \|\int_0^1 F'(x^*)^{-1}\Big(F''(x_m+\tau(x^*-x_m))-F''(x^*)\Big)(1-\tau)d\tau\|\,\|x^*-x_m\|^2 \\
&+\int_0^1 (1-\tau)d\tau\|F'(x^*)^{-1}F''(x^*)\|\,\|x^*-x_m\|^2 \\
&\leq \int_0^1 w\Big((1-\theta)\|x^*-x_m\|\Big)(1-\tau)d\tau\|x^*-x_m\|^2+\frac{1}{2}b_2\|x^*-x_m\|^2. \qquad (3.13)
\end{aligned}$$

□

Remark 3.3 In view of the estimate

$$\begin{aligned}
&F'(x^*)^{-1}\Big(F'(x^*)-F(x)\Big) \\
&=-F'(x^*)\Big[F'(x)-F'(x^*)-F''(x^*)(x-x^*)+F''(x^*)(x-x^*)\Big] \\
&=-F'(x^*)^{-1}\int_0^1\Big(F''(x^*+\tau(x-x^*))-F''(x^*)\Big)d\tau(x-x^*)-F'(x^*)^{-1}F''(x^*)(x-x^*), \quad (3.14)
\end{aligned}$$

We can obtain by hypotheses (B_2) and (B_4) that

$$\begin{aligned}
\|F'(x^*)^{-1}\Big(F'(x)-F'(x^*)\Big)\| &\leq \|\int_0^1 F'(x^*)^{-1}\Big(F''(x^*+\tau(x-x^*))-F''(x^*)\Big)d\tau(x-x^*)\| \\
&+\|F'(x^*)^{-1}F''(x^*)(x-x^*)\| \\
&\leq \int_0^1 w(\tau\|x-x^*\|)d\tau\|x-x^*\|+b_2\|x-x^*\| \qquad (3.15)
\end{aligned}$$

so, we can choose

$$V(t) = \Big(\int_0^1 W(\tau t) d\tau + b_2 \Big) t. \tag{3.16}$$

However, the function V can be smaller and more flexible than its special choice given by (3.16) leading to tighter error bounds and a larger radius of convergence (see also the numerical examples in [2,3]).

References

[1] Amat, S., Busquier, S. and Gutierrez, J.M., Geometric constructions of iterative functions to solve nonlinear equations, *J. Comput. Appl. Math.* 157, 2003, 197-205.

[2] Argyros, I.K., *Convergence and Applications of Newton-type Iterations*, Springer Verlag, Berlin, Germany, 2008.

[3] Argyros, I.K. and Magreñán, A.A., *A Contemporary Study of Iterative Methods*, Academic Press, Elsevier, San Diego, CA, USA, 2017.

[4] Denis, J.E. and Schnabel, R.B., *Numerical methods for unconstrained optimization and nonlinear equation*, SIAM, Philadelphia, 1996.

[5] Deulfhard, P. and Heindl, G., Affine invariant convergence theorems for Newton's method and extensions to related methods, *SIAM J. Numer. Anal.*, 16 (1979) 1 - 10.

[6] Ezquerro, J. A. and Hernández, M.A., A semi-local convergence result for Newton's method under generalized conditions of Kantorovich, *J. Complexity*, 30 (2014) 309 - 324.

[7] Ezquerro, J. A. and Hernández, M.A., Center conditions on high order derivates in the semi-local convergence of Newton's method, *J. Complexity*, 31 (2015) 277 - 292.

[8] Ezquerro, J. A. and Hernández M.A., Generalized differentiability conditions for Newton's method, *IMA J. Numer. Anal.*, 22, 2 (2002) 187 - 205.

[9] Ezquerro, J.A. and Hernández, M.A., Newton's method: an updated Approach of Kantorovich's theory, *Frontiers in Mathematics*, Birkhäuser, Cham Switzerland, (2017).

[10] Ezquerro, J. A. and Hernández M.A., On an application of Newton's method to non-linear operators with w-conditioned second derivative, *BIT*, 42, 3 (2002) 519 - 530.

[11] Ezquerro, J. A., González, D. and Hernández, M.A., Majorizing sequences for Newton's method from initial value problem, *J. Comput. Appl. Math.*, 236 (2012) 2246 - 2258.

[12] Gutiérrez, J. M., A new semi-local convergence theorem for Newton's method, *J. Comput. Appt. Math*, 79 (1997) 131-145.

[13] Gutiérrez, J. M. and Hernández, M.A., Majorizing sequences for Newton's method, *Tamkang J. Math.*, 29, 3 (1998) 199-202.

[14] Gutiérrez, J. M. and Hernández M.A., Newton's method under different Lipschitz conditions, (Numerical analysis and its applications) *Lectures Notes in Comput. Sci.*, 1988 (2000) 368 - 376.

[15] Gutiérrez, J. M. and Hernández, M.A., Newton's method under weak Kantorovich conditions, *IMA J. Numer. Anal.*, 20 (2000) 521 - 532.

[16] Hernández, M.A., Relaxing convergence conditions for Newton's method, *J. Math. Anal. Appl.*, 249 (2000) 463 - 475.

[17] Huang, Z., A note on the Kantorovich theorem for Newton method, *J. Comput. Appl. Math.*, 47 (1993) 211 - 217.

[18] Kantorovich, L.V. and Akilov, G. P., *Functional Analysis*, Pergamon Press, New York, 1982.

[19] Ortega, J. M. and Rheinboldt, W. C., *Iterative solutions of nonlinear equations in several variables*, Academic Press, New York, 1970.

Chapter 12

Convergence of an Eight Order Four Step Method

1. Introduction

The local convergence analysis of method

$$\begin{aligned} y_n &= x_n - F'(x_n)^{-1}F(x_n) \\ z_n &= y_n - 5F'(x_n)^{-1}F(y_n) \\ w_n &= z_n - \frac{16}{5}F'(x_n)^{-1}F(y_n) - \frac{1}{5}F'(x_n)^{-1}F(z_n) \\ x_{n+1} &= w_n - H(u_n)F'(x_n)^{-1}F(w_n), \end{aligned} \tag{1.1}$$

where

$$u_n = I - 5F'(x_n)^{-1}[y_n, x_n; F],$$

$[.,.;F] : D \times D \to L(X,Y)$ is a divided difference of order, $H : L(X,Y) \to L(X,Y)$, and $x_0 \in D$ was studied, when $X = Y = \mathbb{R}^k$. The convergence order was shown to be eight under hypotheses up to the ninth derivative (although these derivatives do not appear in (1.1)) limiting the applicability of the method. As a motivational example, consider $D = \left[-\frac{3}{2}, \frac{1}{2}\right]$ and function F be defined on D by

$$F(x) = \begin{cases} x^3 \ln x^2 + x^5 - x^4, & x \neq 0 \\ 0, & x = 0. \end{cases}$$

Then, clearly $F'''(x)$ is unbounded on D. Hence, the earlier results cannot apply.

In this chapter, we show the local convergence analysis using only the first Fréchet-derivative, and in more general setting of Banach space valued operators. Error estimates and uniqueness results are also provide (not given in earlier works). The order of convergences uses computational order of convergence and approximate order of convergence not requiring high convergence order derivatives. This way, we expand the applicability of method and in the more general setting of a Banach space.

2. Local Convergence

Let us define some parameters and real functions to be utilized in the local convergence analysis of method (1.1) that follows. Let $\varphi_0 : [0,\infty) \to [0,\infty)$ be a continuous and increasing function satisfying $\varphi_0(0) = 0$.
Suppose that equation

$$\varphi_0(t) = 1 \tag{2.1}$$

has a minimal positive solution δ_0. Let $\varphi : [0,\delta_0] \to [0,\infty), \varphi_1 : [0,\delta_0) \to [0,\infty)$, and $\Psi : [0,\delta_0) \to [0,\infty)$ be also continuous and increasing functions satisfying $\varphi(0) = 0$. Define functions λ_i, μ_i, for $i = 1,2,3,4$ on the interval $[0,\delta_0)$ by

$$\lambda_1(t) = \frac{\int_0^1 \varphi\Big((1-\theta)t\Big)d\theta}{1-\varphi_0(t)},$$

$$\lambda_2(t) = \left(1 + \frac{5\int_0^1 \varphi_1\Big(\theta\lambda_1(t)t\Big)d\theta}{1-\varphi_0(t)}\right)\lambda_1(t),$$

$$\lambda_3(t) = \lambda_2(t) + \frac{16}{5}\frac{\int_0^1 \varphi_1\Big(\theta\lambda_1(t)t\Big)d\theta\lambda_1(t)}{1-\varphi_0(t)} + \frac{\int_0^1 \varphi_1\Big(\theta\lambda_2(t)t\Big)d\theta\lambda_2(t)}{5\big(1-\varphi_0(t)\big)},$$

$$\lambda_4(t) = \left(1 + \frac{\Psi(t)\int_0^1 \varphi_1(\theta\lambda_3(t)t)d\theta}{1-\varphi_0(t)}\right)\lambda_3(t),$$

and

$$\mu_i(t) = \lambda_i(t) - 1.$$

We have $\mu_i(0) = -1$ and $\mu_i(t) \to \infty$ as $t \to \delta_0^-$. The intermediate value theorem guarantees that equation $\mu_i(t) = 0$ have solutions in $(0, \delta_0)$. Denote by δ_i the smallest such solutions, respectively. Define a radius of convergence δ by

$$\delta = \min\{\delta_i\}. \tag{2.2}$$

Then, for each $t \in [0, \delta)$

$$0 \leq \lambda_i(t) < 1. \tag{2.3}$$

Define $U(x^*, a) = \{x \in X : \|x - x^*\| < a\}$ for $a > 0$. Moreover, denoted by $\overline{U}(x^*, a)$ the closure of $U(x^*, a)$.

The local convergence analysis is based on the conditions (C):

(C_1) $F : D \subset X \to Y$ is Fréchet-differentiable, and there exists $x^* \in D$ such that $F(x^*)$ and $F'(x^*)^{-1} \in L(Y, X)$. Moreover $[\cdot\ , \cdot\ ; F] : DxD \to L(X, Y)$ is divided difference of order one.

(C_2) There exists a continuous and increasing function $\varphi_0 : [0, \infty) \to [0, \infty)$ satisfying $\varphi_0(0) = 0$ such that for each $x \in D$

$$\|F'(x^*)^{-1}(F'(x) - F'(x^*))\| \leq \varphi_0(\|x - x^*\|).$$

Set $D_0 = D \cap U(x^*, \delta_0)$.

(C_3) There exist continuous and increasing functions $\varphi : [0, \delta_0) \to [0, \infty), \varphi_1 :$ $[0, \delta_0) \to [0, \infty)$ satisfying $\varphi(0) = 0$ such that for each $x, y \in D_0$.

$$\|F'(x^*)^{-1}(F'(x) - F'(y))\| \leq \varphi(\|x - y\|).$$

and

$$\|F'(x^*)^{-1}F'(x)\| \leq \varphi_1(\|x - x^*\|).$$

(C_4) There exist $H : L(X, Y) \to L(X, Y)$, a continuous and increasing function $\Psi : [0, \delta) \to [0, \infty)$ such that for each $u \in L(X, Y)$ and $x \in D_0$.

$$\|H(u)\| \leq \Psi(\|x - x^*\|),$$

where,

$$u(x) = u = I - 5F'(x)^{-1}\Big[x - F'(x)^{-1}F(x), x - F'(x)^{-1}F(x) -$$
$$5F'(x)^{-1}F\big(x - F'(x)^{-1}F(x)\big); F\Big],$$

(C_5) (2.1) holds and $\overline{U}(x^*, \delta) \subseteq D$, where δ is defined in (2.2).

(C_6) There exists $\bar{\delta} \geq \delta$ such that

$$\int_0^1 \varphi_0(\theta\bar{\delta})d\theta < 1.$$

Set $D_1 = D \cap \overline{U}(x^*, \bar{\delta})$.

Next, the conditions (c) together with the preceding notation are used to show the local convergence analysis that follows.

Theorem 2.1 Suppose that the (c) conditions hold. Then, sequence $\{x_n\}$ generated by method (1.1) is well defined, for each $n = 0, 1, 2, \ldots$ remains in $U(x^*, \delta)$ and converges to a unique solution of equation $F(x) = 0$ in D_1 provided that $x_0 \in U(x^*, \delta) - \{x^*\}$. Moreover, the following error bounds hold:

$$\|y_n - x^*\| \leq \lambda_1(\|x_n - x^*\|)\|x_n - x^*\| \leq \|x_n - x^*\| \leq \delta, \tag{2.4}$$

$$\|z_n - x^*\| \leq \lambda_2(\|x_n - x^*\|)\|x_n - x^*\| \leq \|x_n - x^*\|, \tag{2.5}$$

$$\|w_n - x^*\| \leq \lambda_3(\|x_n - x^*\|)\|x_n - x^*\| \leq \|x_n - x^*\|, \tag{2.6}$$

and

$$\|x_{n+1} - x^*\| \leq \lambda_4(\|x_n - x^*\|)\|x_n - x^*\| \leq \|x_n - x^*\|, \tag{2.7}$$

where the functions $\delta_i, i = 1, 2, 3, 4$, set D_1 and radius of convergence are defined previously.

Proof. We shall show estimates (2.4)-(2.7) holds by using mathematical induction. By hypothesis $x_0 \in U(x^*, \delta) - \{x^*\}$, (2.2) and (2.3), we have in turn that

$$\|F'(x^*)^{-1}(F'(x_0) - F'(x^*))\| \leq \varphi_0(\|x_0 - x^*\|) \leq \varphi_0(\delta) < 1. \tag{2.8}$$

The Banach lemma on invertible operators [15], and (2.8) guarantee $F'(x_0)^{-1} \in L(Y,X)$ and

$$\|F'(x_0)^{-1}F'(x^*)\| \le \frac{1}{1-\varphi_0\big(\|x_0-x^*\|\big)}. \tag{2.9}$$

The points y_0, z_0, w_0, x_1 are also well defined by method (1.1) for $n=0$. Using, the first sub step of method (1.1) for $n=0, (C_1), (2.2)$, (2.3)(for i=1), first condition in (C_3) and (2.9), we get in turn that

$$\begin{aligned}\|y_0-x^*\| &= \|x_0-x^*-F'(x_0)^{-1}F(x_0)\| \\ &\le \|F'(x_0)^{-1}F'(x^*)\| \, \|\int_0^1 F'(x^*)^{-1}\Big(F'(x^*+\theta(x_0-x^*))-F'(x_0)\Big)(x_0-x^*)d\theta\| \\ &\le \frac{\int_0^1 \varphi\Big((1-\theta)\|x_0-x^*\|\Big)d\theta\|x_0-x^*\|}{1-\varphi_0(\|x_0-x^*\|)} \\ &= \lambda_1(\|x_0-x^*\|)\|x_0-x^*\| \le \|x_0-x^*\| < \delta, \end{aligned} \tag{2.10}$$

which shows (2.4) holds for $n=0$, and $y_0 \in U(x^*,\delta)$. We can write by (C_1) and the second condition in (C_3) that

$$F(x_0) = F(x_0)-F(x^*) = \int_0^1 F'\big(x^*+\theta(x_0-x^*)\big)d\theta(x_0-x^*)$$

so

$$\|F'(x^*)^{-1}F(x_0)\| \le \int_0^1 \varphi_1\big(\theta\|x_0-x^*\|\big)d\theta\|x_0-x^*\|, \tag{2.11}$$

since, $\|x^*+\theta(x_0-x^*)-x^*\| = \theta\|x_0-x^*\| < \delta$

$$(i.e.,\ x^*+\theta(x_0-x^*) \in U(x^*,\delta) \text{ for each } \theta \in [0,1]).$$

Therefore, by the second sub-step of method (1.1) for $n=0$, (2.2), (2.3)(for

$i = 2$), (2.8)(for $y_0 = x_0$), (2.9) - (2.11), we obtain in turn that

$$\begin{aligned}
\|z_0 - x^*\| &= \|y_0 - x^* - 5F'(x_0)^{-1}F(y_0)\| \\
&\leq \|y_0 - x^*\| + 5\|F'(x_0)^{-1}F'(x^*)\| \, \|F'(x^*)^{-1}F(y_0)\| \\
&\leq \|y_0 - x^*\| + 5\frac{\int_0^1 \varphi_1\left(\theta\|y_0 - x^*\|\right)d\theta\|y_0 - x^*\|}{1 - \varphi_0(\|x_0 - x^*\|)} \\
&\leq \left(1 + 5\frac{\int_0^1 \varphi_1(\theta\lambda_1(\|x_0 - x^*\|)\|x_0 - x^*\|d\theta)}{1 - \varphi_0(\|x_0 - x^*\|)}\right)\lambda_1(\|x_0 - x^*\|)\|x_0 - x^*\| \\
&= \lambda_2(\|x_0 - x^*\|)\|x_0 - x^*\| \leq \|x_0 - x^*\|, \qquad (2.12)
\end{aligned}$$

which shows (2.5) for $n = 0$, and $z_0 \in U(x^*, \delta)$. By the third sub-step of method (1.1) for $n = 0$, (2.2), (2.3) (for $i = 3$), (2.9), (2.10), (2.11) (for $x_0 = z_0$), we have in turn that

$$\begin{aligned}
\|w_0 - x^*\| &= \|z_0 - x^* - \frac{16}{5}F'(x_0)^{-1}F(y_0) - \frac{1}{5}F'(x_0)^{-1}F(z_0)\| \\
&\leq \|z_0 - x^*\| + \frac{16}{5}\|F'(x_0)^{-1}F'(x^*)\| \, \|F'(x^*)^{-1}F(y_0)\| \\
&\quad + \frac{1}{5}\|F'(x_0)^{-1}F(x^*)\| \, \|F'(x^*)^{-1}F(z_0)\| \\
&\leq \lambda_2(\|x_0 - x^*\|)\|x_0 - x^*\| \\
&\quad + \frac{16}{5}\frac{\int_0^1 \varphi\left(\theta\lambda_1(\|x_0 - x^*\|)\|x_0 - x^*\|\right)d\theta\lambda_1(\|x_0 - x^*\|)\|x_0 - x^*\|}{1 - \varphi_0(\|x_0 - x^*\|)} \\
&\quad + \frac{1}{5}\frac{\int_0^1 \varphi_1\left(\theta\lambda_2(\|x_0 - x^*\|)\|x_0 - x^*\|\right)d\theta\lambda_2(\|x_0 - x^*\|)\|x_0 - x^*\|}{1 - \varphi_0(\|x_0 - x^*\|)} \\
&= \lambda_3(\|x_0 - x^*\|)\|x_0 - x^*\| \leq \|x_0 - x^*\|, \qquad (2.13)
\end{aligned}$$

which shows (2.6) for $n = 0$, and $w_0 \in U(x^*, \delta)$.
The last sub-step of method (1.1) for $n = 0$, (2.2), (2.3)(for $i = 4$), (C_4), (2.9),

(2.11(for $x_0 = w_0$), and (2.14), give in turn that

$$\begin{aligned} \|x_1 - x^*\| &= \|w_0 - x^* - H(u_0)F'(x_0)^{-1}F(w_0)\| \\ &\leq \|w_0 - x^*\| + \|H(u_0)\| \, \|F'(x_0)^{-1}F'(x^*)\| \, \|F'(x^*)^{-1}F(w_0)\| \\ &\leq \left(1 + \frac{\Psi(\|x_0 - x^*\|) \int_0^1 \varphi_1(\theta\|w_0 - x^*\|)d\theta}{1 - \varphi_0(\|x_0 - x^*\|)}\right) \|w_0 - x^*\| \\ &\leq \lambda_4(\|x_0 - x^*\|)\|x_0 - x^*\| \leq \|x_0 - x^*\|, \end{aligned} \tag{2.14}$$

which shows (2.7) for $n = 0$ and $x_1 \in U(x^*, \delta)$.
Replace x_0, y_0, z_0, w_0, x_1 by $x_m, y_m, z_m, w_m, x_{m+1}$, respectively in the preceding estimates to obtain (2.4)-(2.7) hold for each $m = 0, 1, 2, \ldots$. In view of the estimates

$$\|x_{m+1} - x^*\| \leq c\|x_m - x^*\| < \delta, \tag{2.15}$$

and

$$c = \lambda_4(\|x_0 - x^*\|) \in [0, 1),$$

we conclude that $\lim_{m\to+\infty} x_m = x^*$ and $x_{m+1} \in U(x^*, \delta)$. Let us define

$$T = \int_0^1 F'(x^* + \theta(y^* - x^*))d\theta,$$

for some $y^* \in D_1$ with $F(y^*) = 0$. Then, using (C_2) and (C_6) we get in turn that

$$\|F'(x^*)(T - F'(x^*))\| \leq \int_0^1 \varphi_0(\theta\|x^* - y^*\|)d\theta \leq \int_0^1 \varphi_0(\theta\bar{\delta})d\theta < 1, \tag{2.16}$$

so $T^{-1} \in L(Y, X)$. Then, for the identity

$$0 = F(y^*) - F(x^*) = T(y^* - x^*), \tag{2.17}$$

we deduce that $x^* = y^*$.

□

We shall show how to find an upper bound on $\|u_n\|$.

(C_7) Suppose that there exists function $\overline{\varphi}_0 : [0,\rho_0) \times [0,\rho_0) \to [0,\infty)$ continuous, and increasing such that for each $x, y \in D_0$

$$\|F'(x^*)^{-1}\Big([y,z;F] - F'(x^*)\Big)\| \le \overline{\varphi}_0(\|y - x^*\|, \|x - x^*\|).$$

Then, we have

$$\begin{aligned}
\|u_n\| &= \|F'(x_n)^{-1}\Big(F'(x_n) - 5[y_n,z_n;F]\Big)\| \\
&\le \|F'(x_n)^{-1}F'(x^*)\|\|F'(x^*)^{-1}\Big(F'(x_n) - F'(x^*)\Big) \\
&\qquad - 5\Big([y_n,z_n;F] - F'(x^*)\Big) - 4F'(x^*)\| \\
&\le \frac{4 + \varphi_0(\|x_n - x^*\|) + \overline{\varphi}_0(\|y_n - x^*\|, \|z_n - x^*\|)}{1 - \varphi_0(\|x_0 - x^*\|)} \\
&\le \varphi_2(\|x_0 - x^*\|),
\end{aligned} \tag{2.18}$$

where

$$\varphi_2(t) = \frac{4 + \varphi_0(t) + \overline{\varphi}_0\big(g_1(t)t, g_2(t)t\big)}{1 - \varphi_0(t)}.$$

Choose for simplicity $H(v) = bv$, for some $b > 0$, then

$$\|H(u)\| \le |b|\varphi_2(t) \text{ for } t \in [0,\rho_0). \tag{2.19}$$

References

[1] Amat, S., Busquier, S. and Gutierrez, J.M., Geometric constructions of iterative functions to solve nonlinear equations, *J. Comput. Appl. Math.* 157, 2003, 197-205.

[2] Argyros, I.K., *Convergence and Applications of Newton-type Iterations*, Springer Verlag, Berlin, Germany, 2008.

[3] Argyros, I.K. and Chen, D., Results on the Chebsyhev method in Banach spaces, *Proyecciones* 12, 1993, 119-128.

[4] Argyros, I.K. and Magreñán, A.A., *A Contemporary Study of Iterative Methods*, Academic Press, Elsevier, San Diego, CA, USA, 2017.

[5] Argyros, I.K., Chen, D. and Qian, Q., A convergence analysis for rational methods with a parameter in Banach space, *Pure Math. Appl.* 5, 1994, 59-73.

[6] Argyros, I.K., On an improved unified convergence analysis for a certain class of Euler-Halley type methods, *J. Korean Soc. Math. Educ. Ser. B Pure Appl. Math.* 13, 2006, 207-215.

[7] Chen, D., Argyros, I.K. and Qian, Q., A local convergence theorem for the Supper-Halley method in a Banach space, *Appl. Math. Lett.* 7, 1994, 49-52.

[8] Ezquerro, J.A. and Hernandez, M.A., Halley's method for operators with unbounded second derivative, *Appl. Numer. Math.* 57, 2007, 354-360.

[9] Ezquerro, J.A. and Hernandez, M.A., New Kantorovich-type conditions for Halley's method, *Appl. Number. Anal. Comput. Math.* 2, 2005, 70-77.

[10] Ezquerro, J.A. and Hernandez, M.A., On the R-order of the Halley method, *J. Math. Anal. Appl.* 303, 2005, 591-601.

[11] Gutierrez, J.M. and Hernandez, M.A., A family of Chebyshev-Halley type methods in Banach space, *Bull. Aust. Math. Soc.* 55, 1997, 113-130.

[12] Gutierrez, J.M. and Hernandez, M.A., An acceleration of Newton's method: Super-Halley method, *Appl. Math. Comput.* 117, 2001, 223-239.

[13] Hernandez, M.A. and Salanova, M.A., A family of Chebyshev-Halley type methods. *Int. J. Comput. Math.* 47, 1993, 59-63.

[14] Huang, Z., On a family of Chebyshev-Halley type methods in Banach space under weaker Smale condition. *Number. Math. JCU* 9, 2000, 37-44.

[15] Kantorovich, L.V. and Akilov, G.P., *Functional Analysis in Normed Spaces*, Pergamon Press, New York, 1964.

Chapter 13

A Fast Three Step Method

1. Introduction

The local convergence analysis of method

$$\begin{aligned} y_n &= x_n - F'(x_n)^{-1}F(x_n) \\ z_n &= h(x_n, y_n) \\ x_{n+1} &= z_n - F'(y_n)^{-1}F(z_n), \end{aligned} \tag{1.1}$$

where $x_0 \in D$ was studied, when $X = Y = \mathbb{R}^k$. The convergence order was shown to be p under hypotheses up to the $p+1$ derivative (although these derivatives do not appear in (1.1)) limiting the applicability of the method. As a motivational example, consider $D = \left[-\frac{3}{2}, \frac{1}{2}\right]$ and function F be defined on D by

$$F(x) = \begin{cases} x^3 \ln x^2 + x^5 - x^4, & x \neq 0 \\ 0, & x = 0. \end{cases}$$

Then, clearly $F'''(x)$ is unbounded on D. Hence, the earlier results cannot apply.

In this chapter, we show the local convergence analysis using only the first Fréchet-derivative, and in more general setting of Banach space valued operators. Error estimates and uniqueness results are also provide (not given in earlier works). The order of convergences uses computational order of convergence and approximate order of convergence not requiring high convergence order derivatives. This way, we expand the applicability of method and in the more general setting of a Banach space.

2. Local Convergence

We introduce some parameters and real functions needed for the local convergence analysis of method (1.1). Let $\varphi_0 : [0,\infty) \to [0,\infty)$ be a continuous and increasing function with $\varphi_0(0) = 0$.
Suppose that equation

$$\varphi_0(t) = 1 \tag{2.1}$$

has positive solutions. Denote by r_0 the smallest such solution. Let also $\varphi : [0,r_0) \to [0,\infty)$, $\varphi_1 : [0,r_0) \to [0,\infty)$ and $\Psi : [0,r_0) \to [0,\infty)$ be continuous and increasing functions with $\varphi(0) = \Psi(0) = 0$.

Define functions λ_1, μ_1 and μ_2 on the interval $[0,r_0)$ by

$$\lambda_1(t) = \frac{\int_0^1 \varphi\Big((1-\theta)t\Big)d\theta}{1-\varphi_0(t)}$$

$$\mu_1(t) = \lambda_1(t) - 1,$$

and,

$$\mu_2(t) = \Psi(t) - 1.$$

We have that $\mu_1(0) = -1$ and $\mu_1(t) \to \infty$ as $t \to r_0^{-1}$. It follows by the intermediate value theorem that equation $\mu_1(t) = 0$ has solutions in the interval $(0,r_0)$. Denote by r_1 the smallest such zero.
Suppose that equation

$$\Psi(t) = 1 \tag{2.2}$$

has solutions in $(0,r_0)$. Denote by r_2 the smallest such solution.
Suppose that equation

$$\varphi_0\big(\lambda_1(t)t\big) = 1 \tag{2.3}$$

has solutions in $(0, r_0)$. Denote by $\overline{r}_3$ the smallest such solution. Define functions $\lambda_3(t)$ and μ_3 on the interval $[0, \overline{r}_3)$ by

$$\lambda_3(t) = \left(1 + \frac{\int_0^1 \varphi_1\left(\theta\Psi(t)t\right)d\theta}{1 - \varphi_0\left(\lambda_1(t)t\right)}\right)\Psi(t)$$

and

$$\mu_3(t) = \lambda_3(t) - 1.$$

We obtain $\mu_3(0) = -1$ and $\mu_3(t) \to \infty$ as $t \to \overline{r}_3^{\,-}$. Denote by r_3 the smallest such solution of equation $\mu_3(t) = 0$ in $(0, \overline{r}_3)$. Define the radius o convergence r by

$$r = \min\{r_1\}, i = 1, 2, 3. \tag{2.4}$$

It follows that for each $t \in [0, r)$

$$0 \leq \varphi_0(t) < 1, \tag{2.5}$$

$$0 \leq \varphi_0\left(\lambda_1(t)t\right) < 1, \tag{2.6}$$

and

$$0 \leq \lambda_i(t) < 1. \tag{2.7}$$

We denote by $U(u, \delta)$ and $\overline{U}(u, \delta)$ be the open and closed balls in X, respectively with center $u \in X$ and of radius $\delta > 0$.

The hypothesis (H) shall be used to show the local convergence of method (1.2):

(H_1) $F : \Omega \subseteq X \to Y$ is a continuously Fréchet-differentiable operator and there exists $x^* \in \Omega$ such that $F(x^*) = 0$ and $F'(x^*)^{-1} \in L(Y, X)$.

(H_2) There exists a continuous and increasing function $\varphi_0 : [0, \infty) \to [0, \infty)$ with $\varphi_0(0) = 0$ such that

$$\|F'(x^*)^{-1}\left(F'(x) - F'(x^*)\right)\| \leq \varphi_0(\|x - x^*\|).$$

Set $\Omega_0 = \Omega \cap U(x^*, r_0)$, where r_0 is given in (2.1).

(H_3) There exist continuous and increasing functions $\varphi : [0, r_0) \to [0, \infty)$, $\varphi_1 : [0, r_0) \to [0, \infty)$, $\Psi : [0, r_0) \to [0, \infty)$ and continuous operator $h : \Omega_0 \times \Omega_0 \to X$ such that for each $x, y \in \Omega_0$ such that for each $x, y \in \Omega_0$

$$\|F'(x^*)^{-1}\Big(F'(x) - F'(y)\Big)\| \leq \varphi(\|x - y\|),$$

$$\|F'(x^*)^{-1}F'(x)\| \leq \varphi_1(\|x - x^*\|),$$

and

$$\|h(x, v) - x^*\| \leq \Psi(\|x - x^*\|)\|x - x^*\|$$

for $v = x - F'(x)^{-1}F(x)$.

(H_4) $\overline{U}(x^*, r) \subseteq D$ and (2.1)-(2.3) hold, where r is given in (2.4).

(H_4) There exists $r^* \geq r$ such that

$$\int_0^1 \varphi_0(\theta r^*)d\theta < 1.$$

Set $\Omega_1 = \Omega \cap \overline{U}(x^*, r^*)$.

Next, the hypotheses (H) as well as the preceding notation are used to show the local convergence method of (1.2).

Theorem 2.1 Suppose that the conditions (H) hold and $x_0 \in U(x^*, r) - \{x^*\}$. Then, iteration $\{x_n\}$ generated by the method (1.2) exists, remain in $U(x^*, r)$ for each $n = 01, 2, \ldots$ and converges to x^*, which is the only solution of equation $F(x) = 0$ in Ω_1. Moreover, the following error bounds hold

$$\|y_n - x^*\| \leq \lambda_1(\|x_n - x^*\|) \leq \|x_n - x^*\| < r, \tag{2.8}$$

$$\|z_n - x^*\| \leq \Psi(\|x_n - x^*\|)\|x_n - x^*\| \leq \|x_n - x^*\| \tag{2.9}$$

and

$$\|x_{n+1} - x^*\| \leq \lambda_3(\|x_n - x^*\|)\|x_n - x^*\| \leq \|x_n - x^*\|, \tag{2.10}$$

where the radius of convergence r is given by (2.4) and functions $\lambda_1, \lambda_3, \Psi$ are defined previously.

Proof. We shall use mathematical induction to show error bounds (2.8)-(2.10). Using (2.1), (2.5), (H_1), and hypothesis $x_0 \in U(x^*, r) - \{x^*\}$, we obtain in turn that

$$\|F'(x^*)^{-1}\Big(F'(x_0) - F'(x^*)\Big)\| \leq \varphi_0(\|x_0 - x^*\|) \leq \varphi_0(r) < 1, \tag{2.11}$$

so by the Banach lemma on invertible operators [15], $F'(x_0)^{-1} \in L(Y,X)$ and

$$\|F'(x_0)^{-1}F'(x^*)\| \leq \frac{1}{1 - \varphi_0(\|x_0 - x^*\|)}. \tag{2.12}$$

The iterate y_0 is then well defined by the first sub-step of method (1.1) for $n = 0$. We have that

$$y_0 - x^* = x_0 - x^* - F'(x_0)^{-1}F(x_0) \tag{2.13}$$

Then, by (2.1), (2.4), (2.7)(for $i = 1$), the first hypothesis in (H_3), (2.12) and (2.13), we get in turn that

$$\|y_0 - x^*\| \leq \|F'(x_0)^{-1}F'(x^*)\| \, \|\int_0^1 F'(x^*)^{-1}\Big(F'(x^* + \theta(x_0 - x^*)) - F'(x_0)\Big)(x_0 - x^*)d\theta\|$$

$$\leq \frac{\int_0^1 \varphi((1-\theta)\|x_0 - x^*\|)d\theta\|x_0 - x^*\|}{1 - \varphi_0(\|x_0 - x^*\|)}$$

$$= \lambda_1(\|x_0 - x^*\|)\|x_0 - x^*\| \leq \|x_0 - x^*\| < r. \tag{2.14}$$

which shows (2.8) for $n = 0$ and $y_0 \in U(x^*, r)$.

By the second sub-step of method (1.2), (2.2), (2.4), and the hypothesis in (H_3), we get that

$$\|z_0 - x^*\| = \|h(x_0, y_0) - x^*\| \leq \Psi(\|x_0 - x^*\|)\|x_0 - x^*\| \leq \|x_0 - x^*\|, \tag{2.15}$$

which shows (2.9) for $n = 0$ and $z_0 \in U(x^*, r)$. Then, in view of (2.4), (2.6), and (2.12) for $x_0 = y_0, F'(y_0)^{-1} \in L(Y,X)$ and

$$\|F'(y_0)^{-1}F'(x^*)\| \leq \frac{1}{1 - \varphi_0(\|y_0 - x^*\|)}$$

$$\leq \frac{1}{1 - \varphi_0\big(\lambda_1(\|x_0 - x^*\|)\|x_0 - x^*\|\big)}. \tag{2.16}$$

It also follows that iterates x_1 is well defined by the third sub-step of method (1.1) for $n = 0$. We can write by H_1 that

$$F(z_0) = F(z_0) - F(x^*) = \int_0^1 F'(x^* + \theta(z_0 - x^*))d\theta(z_0 - x^*). \qquad (2.17)$$

Notice that

$$\|x^* + \theta(z_0 - x^*) - x^*\| = \theta\|z_0 - x^*\| < r,$$

so

$$x^* + \theta(z_0 - x^*) \in U(x^*, r) \text{ for each } \theta \in [0, 1)$$

Next, by the second hypothesis in $(H_3), (H_1)$, (2.15), and (2.17), we have that

$$\begin{aligned}\|F'(x^*)^{-1}F(z_0)\| &\leq \int_0^1 \varphi_1(\theta\|z_0 - x^*\|)d\theta\|z_0 - x^*\| \\ &\leq \int_0^1 \varphi_1(\theta\Psi(\|x_0 - x^*\|)\|x_0 - x^*\|)d\theta\Psi(\|x_0 - x^*\|) \qquad (2.18)\end{aligned}$$

Furthermore, from (2.4), (2.7) (for $i = 3$), the third sub-step of method (1.1), (2.16), and (2.18), we obtain in turn that

$$\begin{aligned}\|x_1 - x^*\| &\leq \|z_0 - x^*\| + \|F'(y_0)^{-1}F'(x^*)\|\|F'(x^*)^{-1}F(z_0)\| \\ &\leq \left(1 + \frac{\int_0^1 \varphi_1(\theta\|z_0 - x^*\|)d\theta}{1 - \varphi_0(\|x_0\, x^*\|)}\right)\|z_0 - x^*\| \\ &\leq \lambda_3(\|x_0 - x^*\|)\|x_0 - x^*\| \leq \|x_0 - x^*\|, \qquad (2.19)\end{aligned}$$

which shows (2.10) for $n = 0$ and $x_1 \in U(x^*, r)$.
The induction for (2.8) - (2.10) is finished, if x_0, y_0, z_0, x_1 are replaced by x_m, y_m, z_m, x_{m+1} in the preceding estimates. Then, from the the estimate

$$\|x_{m+1} - x^*\| \leq c\|x_m - x^*\| < r, \qquad (2.20)$$

we deduce that,

$$\lim_{m\to+\infty} x_m = x^* \text{ and } x_{m+1} \in U(x^*, r),$$

where

$$c = \lambda_3(\|x_0 - x^*\|) \in [0,1),$$

Finally, the uniqueness part is shown by letting,

$$T = \int_0^1 F'(x^* + \theta(y^* - x^*))d\theta \text{ for some } y^* \in \Omega_1 \text{ with } F(y^*) = 0.$$

In view of (H_2) and (H_5), we have in turn that

$$\begin{aligned} \|F'(x^*)^{-1}\Big(T - F'(x^*)\Big)\| &\le \int_0^1 \varphi_0(\theta\|x^* - y^*\|)d\theta \\ &\le \int_0^1 \varphi_0(\theta r^*)d\theta < 1, \end{aligned} \tag{2.21}$$

so $T^{-1} \in L(Y,X)$. Then, by the identity

$$0 = F(y^*) - F(x^*) = T(y^* - x^*), \tag{2.22}$$

we conclude that $x^* = y^*$.

□

Applic 2.2 Frontini - Sorman (Fifth Order)

$$\begin{aligned} z_n &= h_1(x_n, y_n), \\ h_1(x,y) &= x - 2\Big(F'(y) + F'(x)\Big)^{-1} F(x). \end{aligned}$$

We can write

$$\begin{aligned}
h_1(x_n,y_n)-x^* &= x_n-x^*-2\Big(F'(y_n)+F'(x_n)\Big)^{-1}F(x_n)\\
&= x_n-x^*-F'(x_n)^{-1}F(x_n)\\
&\quad+\Big[F'(x_n)^{-1}-2\Big(F'(y_n)+F'(x_n)\Big)^{-1}\Big]F(x_n)\\
&= x_n-x^*-F'(x_n)^{-1}F(x_n)+F'(x_n)^{-1}\Big[I\\
&\quad-2F'(x_n)\Big(F'(y_n)+F'(x_n)\Big)^{-1}\Big]F(x_n)\\
&= x_n-x^*-F'(x_n)^{-1}F(x_n)+F'(x_n)^{-1}\Big[(F'(y_n)\\
&\quad+F'(x_n))-2F'(x_n)\Big]\Big(F'(y_n)+F'(x_n)\Big)^{-1}F(x_n)\\
&= x_n-x^*-F'(x_n)^{-1}F(x_n)+F'(x_n)^{-1}\Big[(F'(y_n)-F'(x^*))\\
&\quad+\Big(F'(x^*)-F'(x_n)\Big)\Big]\Big(F'(y_n)+F'(x_n)\Big)^{-1}F(x_n). \qquad (2.23)
\end{aligned}$$

Using the proof of Theorem 2.1 and (2.23), we get in turn that

$$\|h_1(x_n,y_n)-x^*\|\le\lambda_1(b)b+\frac{\Big(\varphi_0(b)+\varphi_0(\lambda_1(b)b)\Big)\int_0^1\varphi_1(\theta b)d\theta b}{2\Big(1-\varphi_0(b)\Big)\Big(1-\frac{1}{2}(\varphi_0(b)+\varphi_0(\lambda_1(b)b))\Big)},$$

where

$$b=\|x_n-x^*\|.$$

Hence, we can define

$$\Psi(t)=\lambda_1(t)+\frac{\Big(\varphi_0(t)+\varphi_0(\lambda_1(t)t)\Big)\int_0^1\varphi_1(\theta t)d\theta}{2\Big(1-\varphi_0(t)\Big)\Big(1-\frac{1}{2}(\varphi_0(t)+\varphi_0(\lambda_1(t)t))\Big)} \qquad (2.24)$$

Applic 2.3 Cordero et al. [13] (Sixth Order)

$$\begin{aligned}
z_n &= h_2(x,y_n),\\
h_2(x,y) &= y-F'(x)^{-1}[2I-F'(y)F'(x)^{-1}]F(y).
\end{aligned}$$

We can write

$$\begin{aligned} h_1(x_n,y_n)-x^* &= y_n-x^*-F'(x_n)^{-1}\Big[2F'(x_n)-F'(y_n)\Big]F'(x_n)^{-1}F(y_n) \\ &= y_n-x^*-F'(x_n)^{-1}\Big[2\Big(F'(x_n)-F'(x^*)\Big) \\ &\quad -\Big(F'(y_n)-F'(x^*)\Big)+F'(x^*)\Big]F'(x_n)^{-1}F(y_n). \end{aligned}$$

Hence, we get

$$\begin{aligned} &\|h_1(x_n,y_n)-x^*\| \le \|y_n-x^*\| \\ &\quad + \frac{[2\varphi_0(\|x_n-x^*\|)+\varphi_0(\|y_n-x^*\|)+1]\int_0^1 \varphi_1\big(\theta\|y_n-x^*\|\big)d\theta\|y_n-x^*\|}{\Big(1-\varphi_0(\|x_n-x^*\|)\Big)^2}. \end{aligned}$$

Therefore, we can choose

$$\Psi(t)=\lambda_1(t)+\frac{\Big(1+2\varphi_0(t)+\varphi_0(\lambda_1(t)t)\Big)\int_0^1 \varphi_1\Big(\theta\lambda_1(t)t\Big)d\theta\lambda_1(t)}{\Big(1-\varphi_0(t)\Big)^2}.$$

References

[1] Amat, S., Busquier, S. and Gutierrez, J.M., Geometric constructions of iterative functions to solve nonlinear equations, *J. Comput. Appl. Math.* 157, 2003, 197-205.

[2] Argyros, I.K., *Convergence and Applications of Newton-type Iterations*, Springer Verlag, Berlin, Germany, 2008.

[3] Argyros, I.K. and Chen, D., Results on the Chebsyhev method in Banach spaces, *Proyecciones* 12, 1993, 119-128.

[4] Argyros, I.K. and Magreñán, A.A., *A Contemporary Study of Iterative Methods*, Academic Press, Elsevier, San Diego, CA, USA, 2017.

[5] Argyros, I.K., Chen, D. and Qian, Q., A convergence analysis for rational methods with a parameter in Banach space, *Pure Math. Appl.* 5, 1994, 59-73.

[6] Argyros, I.K., On an improved unified convergence analysis for a certain class of Euler-Halley type methods, *J. Korean Soc. Math. Educ. Ser. B Pure Appl. Math.* 13, 2006, 207-215.

[7] Chen, D., Argyros, I.K. and Qian, Q., A local convergence theorem for the Supper-Halley method in a Banach space, *Appl. Math. Lett.* 7, 1994, 49-52.

[8] Ezquerro, J.A. and Hernandez, M.A., Halley's method for operators with unbounded second derivative, *Appl. Numer. Math.* 57, 2007, 354-360.

[9] Ezquerro, J.A. and Hernandez, M.A., New Kantorovich-type conditions for Halley's method, *Appl. Number. Anal. Comput. Math.* 2, 2005, 70-77.

[10] Ezquerro, J.A. and Hernandez, M.A., On the R-order of the Halley method, *J. Math. Anal. Appl.* 303, 2005, 591-601.

[11] Gutierrez, J.M. and Hernandez, M.A., A family of Chebyshev-Halley type methods in Banach space, *Bull. Aust. Math. Soc.* 55, 1997, 113-130.

[12] Gutierrez, J.M. and Hernandez, M.A., An acceleration of Newton's method: Super-Halley method, *Appl. Math. Comput.* 117, 2001, 223-239.

[13] Hernandez, M.A. and Salanova, M.A., A family of Chebyshev-Halley type methods. *Int. J. Comput. Math.* 47, 1993, 59-63.

[14] Huang, Z., On a family of Chebyshev-Halley type methods in Banach space under weaker Smale condition. *Number. Math. JCU* 9, 2000, 37-44.

[15] Kantorovich, L.V. and Akilov, G.P., *Functional Analysis in Normed Spaces*, Pergamon Press, New York, 1964.

Chapter 14

Convergence of a Method Using Derivatives and Divided Differences

1. Introduction

The local convergence analysis of method

$$\begin{aligned} y_n &= x_n - \frac{1}{2}F'(x_n)^{-1}F(x_n) \\ z_n &= y_n - \left(2[x_n, y_n; F]^{-1} - F'(x_n)^{-1}\right)F(y_n) \\ x_{n+1} &= z_n - \left(3I - 2F'(x_n)^{-1}[x_n, y_n; F]\right)F'(x_n)^{-1}F(z_n), \end{aligned} \tag{1.1}$$

where

$$[.,.] : D \times D \to L(X,Y)$$

is a divided difference of order one, and $x_0 \in D$ was studied, when $X = Y = \mathbb{R}^k$. The convergence order was shown to be six under hypotheses up to the seventh derivative (although these derivatives do not appear in (1.1)) limiting the applicability of the method. As a motivational example, consider $D = \left[-\frac{3}{2}, \frac{1}{2}\right]$

and function F be defined on D by

$$F(x) = \begin{cases} x^3 \ln x^2 + x^5 - x^4, & x \neq 0 \\ 0, & x = 0. \end{cases}$$

Then, clearly $F'''(x)$ is unbounded on D. Hence, the earlier results cannot apply.

In this chapter, we show the local convergence analysis using only the first Fréchet-derivative, and in more general setting of Banach space valued operators. Error estimates and uniqueness results are also provide (not given in earlier works). The order of convergences uses computational order of convergence and approximate order of convergence not requiring high convergence order derivatives. This way, we expand the applicability of method and in the more general setting of a Banach space.

2. Local Convergence

We shall utilize some real parameters and functions to show the local convergence analysis in this section. Let $\varphi_0 : [0,\infty) \to [0,\infty)$ be a continuous and increasing function satisfying $\varphi_0(0) = 0$. Suppose that equation

$$\varphi_0(t) = 1 \tag{2.1}$$

has a minimal positive solution R_0.
Let $\varphi, \varphi_1, \varphi_2, \Psi : [0,R_0) \to [0,\infty)$ be also continuous and increasing functions satisfying $\varphi(0) = \varphi_2(0) = \Psi(0) = 0$. Define functions λ_1, μ_1 on the interval $[0,R_0)$ by

$$\lambda_1(t) = \frac{\int_0^1 \varphi_0\Big((1-\theta)t\Big)d\theta}{1-\varphi_0(t)},$$

and

$$\mu(t) = \lambda_1(t) - 1.$$

We have $\mu_1(0) = -1$ and $\mu_1(t) \to \infty$ as $t \to R_0^-$. The intermediate value theorem guarantees that equation $\mu_1(t) = 0$ have solutions in $(0,R_0)$. Denote by R_1 the

smallest such solutions.
Suppose that equation

$$\varphi_0\left(\lambda_1(t)t\right) = 1 \tag{2.2}$$

has a minimal solution $\overline{R}_1$.
Suppose that equation

$$\Psi(t) = 1 \tag{2.3}$$

has a minimal positive solution R_Ψ in $(0, \min\{R_0, R_1\}$. Set $\rho = \min\{R_0, R_1\}$ Moreover, define functions λ_2 and μ_2 on $[0, \rho)$ by

$$\lambda_2(t) = \frac{\int_0^1 \varphi\left((1-\theta)\lambda_1(t)t\right)\lambda_1(t)}{1-\varphi_0(\lambda_1(t)t)} + \frac{\left(2\varphi_2(t)+\varphi_0(t)+\varphi_0(\lambda_1(t)t)\right)\int_0^1 \varphi_1\left(\theta\lambda_1(t)t\right)d\theta\lambda_1(t)}{\left(1-\varphi_0(\lambda_1(t)t)\right)(1-\Psi(t))}$$

and

$$\mu_2(t) = \lambda_2(t) - 1.$$

We get $\mu_2(0) = -1$ and $\mu_2(t) \to \infty$ as $t \to \rho^-$. Denote by R_2 the smallest solution of equation $\mu_2(t) = 0$ in $(0, \rho)$.
Furthermore, define functions λ_3 and μ_3 on $[0, \rho)$ by

$$\lambda_3(t) = \left[1 + \frac{\left(3\varphi_0(t)+\varphi_2(t)+2\right)\int_0^1 \varphi_1(\theta\lambda_2(t)t)d\theta}{(1-\varphi_0(t))^2}\right]\lambda_2(t)$$

and

$$\mu_3(t) = \lambda_3(t) - 1.$$

We obtain $\mu_3(0) = -1$ and $\mu_3(t) \to \infty$ as $t \to \rho^-$. Denote by R_3 the smallest solution of equation $\mu_3(t) = 0$ in $(0, \rho)$. Finally, define a radius of convergence R by

$$R = \min\{R_i\}, i = 1, 2, 3. \tag{2.4}$$

Then, the following hold for each $t \in [0, R)$:

$$0 \le \varphi_0(t) < 1, \tag{2.5}$$

$$0 \le \varphi_0(\lambda_1(t)t) < 1, \tag{2.6}$$

and

$$0 \leq \varphi_i(t) < 1. \tag{2.7}$$

From now on, we denote $U(x,a)$, $\overline{U}(x,a)$ the open and closed balls in X respectively with center $x \in X$ and of radius $a > 0$.

We shall use the conditions (A) in the local convergence analysis:

(A_1) $F : D \subset X \to Y$ is a continuously Fréchet-differentiable, divided difference, $[.,.;F] : D \times D \to L(Y,X)$ of order one, and there exists $x^* \in D$ such that $F(x^*) = 0$ and

$$F'(x^*)^{-1} \in L(Y,X).$$

(A_2) There exists function $\varphi_0 : [0,\infty) \to [0,\infty)$ continuous and increasing satisfying $\varphi_0(0) = 0$ such that for each $x \in D$

$$\|F'(x^*)^{-1}\Big(F'(x) - F'(x^*)\Big)\| \leq \varphi_0(\|x - x^*\|).$$

Set $D_0 = D \cap U(x^*, R_0)$, where R_0 is defined in (2.1).

(A_3) There exist functions $\varphi, \varphi_1, \varphi_2, \Psi : [0,R_0) \to [0,\infty)$ continuous, increasing, satisfying $\varphi(0) = \varphi_2(0) = \Psi(0) = 0$ such that for each $x, y \in D_0$

$$\|F'(x^*)^{-1}\Big(F'(x) - F'(y)\Big)\| \leq \varphi(\|x - y\|),$$

$$\|F'(x^*)^{-1}F'(x)\| \leq \varphi_1(\|x - x^*\|)$$

and

$$\|F'(x^*)^{-1}\Big([x,y;F] - F'(x^*)\Big)\| \leq \varphi_2(\|x - x^*\|)$$

for $y = x - F'(x)^{-1}F(x)$.

(A_4) $\overline{U}(x^*, R) \subseteq D$ and (2.1), (2.2), and (2.3) hold.

(A_5) There exists $R^* \geq R$ such that

$$\int_0^1 \varphi_0(\theta R^*)d\theta < 1.$$

Set $D_1 = D \cap \overline{U}(x^*, R^*)$.

Next, we present the local convergence of method (1.1) based on the preceding formulae and conditions (A).

Theorem 2.1 Suppose that the conditions (A) hold. Then, iteration $\{x_n\}$ initiated $x_0 \in U(x^*,R)-\{x^*\}$ and produced by method (1.1) is defined, stays in $U(x^*,R)$ for each $n=0,1,2,\ldots$, and converges to a unique solution of equation $F(x)=0$ in D_1 so that

$$\|y_n-x^*\| \leq \lambda_1(\|x_n-x^*\|)\|x_n-x^*\| \leq \|x_n-x^*\| < R, \tag{2.8}$$

$$\|z_n-x^*\| \leq \lambda_2(\|x_n-x^*\|)\|x_n-x^*\| \leq \|x_n-x*\|, \tag{2.9}$$

and

$$\|x_{n+1}-x^*\| \leq \lambda_3(\|x_n-x^*\|)\|x_n-x^*\| \leq \|x_n-x*\|. \tag{2.10}$$

where functions $\lambda_i, i=1,2,3$ set D_1 and R are defined previously.

Proof. We prove estimates (2.8) - (2.10) using the method of mathematical induction. By condition $x_0 \in U(x^*,R)-\{x^*\}$, (2.4), (2.5) and (A_2) we get in turn that

$$\|F'(x^*)^{-1}\Big(F'(x_0)-F'(x^*)\Big)\| \leq \varphi_0(\|x_0-x^*\|) \leq \varphi_0(R) < 1, \tag{2.11}$$

so that by the Banach lemma on invertible operators [15]

$$F'(x_0)^{-1} \in L(Y,X),$$

and

$$\|F'(x_0)^{-1}F'(x^*)\| \leq \frac{1}{1-\varphi_0(\|x_0-x^*\|)}. \tag{2.12}$$

It also follows that y_0 is well defined by the first sub-step of method (1.1) for $n=0$.

Using the first sub-step of method (1.1) for $n=0$, (2.4), (2.7) (for$i=1$) and the first condition in (A_3), we have in turn

$$\|y_0 - x^*\| \leq \|F'(x_0)^{-1}F'(x^*)\| \| \int_0^1 F'(x^*)^{-1}\Big(F'(x^* + \theta(x_0 - x^*)) - F'(x_0)\Big) \times$$
$$(x_0 - x^*)d\theta\|$$
$$\leq \frac{\int_0^1 \varphi\Big((1-\theta)\|x_0 - x^*\|\Big)d\theta\|x_0 - x^*\|}{1 - \varphi_0(\|x_0 - x^*\|)}$$
$$\leq \lambda_1(\|x_0 - x^*\|)\|x_0 - x^*\| \leq \|x_0 - x^*\| < R, \tag{2.13}$$

which shows (2.8) for $n = 0$ and $y_0 \in U(x^*, R)$.
We write by (A_1),

$$F(y_0) = F(y_0) - F(x^*) = \int_0^1 F'\Big(x^* + \theta(y_0 - x^*)\Big)d\theta(y_0 - x^*) \tag{2.14}$$

Notice $\|x^* + \theta(y_0 - x^*) - x^*\| = \theta\|y_0 - x^*\| < R$, so $x^* + \theta(y_0 - x^*) \in U(x^*, R)$ for each $\theta \in [0, 1]$.
We must show that $A_0 = 2[x_0, y_0; F] - F'(x_0)$ is invertible. In view of (2.4), (2.6), (2.13) and the third condition in (A_3), we obtain in turn that

$$\|F'(x^*)^{-1}\Big(A_0 - F'(x^*)\Big)\| \leq 2\|F'(x^*)^{-1}\Big([x_0, y_0; F] - F'(x^*)\Big)\|$$
$$+ \|F'(x^*)^{-1}\Big(F'(x_0) - F'(x^*)\Big)\|$$
$$\leq 2\varphi_2(\|x_0 - x^*\|) + \varphi_0(\|x_0 - x^*\|)$$
$$= \Psi(\|x_0 - x^*\|) < \Psi(R) < 1,$$

so $A_0^{-1} \in L(Y, X)$,

$$\|A_0^{-1}F'(x^*)\| \leq \frac{1}{1 - \Psi(\|x_0 - x^*\|)}, \tag{2.15}$$

and z_0 is well defined by the second sub-step of method (1.1) for $n = 0$.

We can write:

$$\begin{aligned}
z_0 - x^* &= y_0 - x^* - F'(y_0)^{-1}F(y_0) \\
&\quad + \left[F'(y_0)^{-1} - \left(2[x_0, y_0; F] - F'(x_0)\right)^{-1}\right]F(y_0) \\
&= y_0 - x^* - F'(y_0)^{-1}F(y_0) \qquad (2.16)\\
&\quad + F'(y_0)^{-1}\Big[2\left([x_0, y_0; F] - F'(x^*)\right) + \left(F'(x^*) - F'(x_0)\right) \\
&\quad + \left(F'(x^*) - F'(y_0)\right)\Big]A_0^{-1}F(y_0).
\end{aligned}$$

Then, by (2.4), (2.7) (for $i = 2$), (2.12) (for $y_0 = x_0$), and (2.13) - (2.16), we have in turn that

$$\begin{aligned}
\|z_0 - x^*\| &\leq \frac{\int_0^1 \varphi\Big((1-\theta)\|y_0 - x^*\|\Big)d\theta\|y_0 - x^*\|}{1 - \varphi_0(\|y_0 - x^*\|)} \\
&+ \frac{\Big(2\varphi_2(\|x_0 - x^*\|) + \varphi_0(\|x_0 - x^*\|) + \varphi_0(\|y_0 - x^*\|)\Big)\int_0^1 \varphi_1(\theta\|y_0 - x^*\|)d\theta\|y_0 - x^*\|}{\Big(1 - \varphi_0(\|y_0 - x^*\|)(1 - \Psi(\|x_0 - x^*\|))\Big)} \\
&\leq \lambda_2(\|x_0 - x^*\|)\|x_0 - x^*\| \leq \|x_0 - x^*\|, \qquad (2.17)
\end{aligned}$$

which shows (2.9) for $n = 0$ and $z_0 \in U(x^*, R)$. Notice that x_1 is well defined by (2.12) and the third sub-step of method (1.1) for $n = 0$, from which we can also write:

$$\begin{aligned}
x_1 - x^* &= z_0 - x^* - F'(x_0)^{-1}\Big[3\Big(F'(x_0) - F'(x^*)\Big) \\
&\quad - \Big([x_0.y_0; F] - F'(x^*)\Big) + 2F'(x^*)\Big]F'(x_0)^{-1}F(z_0) \qquad (2.18)
\end{aligned}$$

Then, by (2.4), (2.7) (for $i = 3$), (2.12), (2.13), (2.14) (for $y_0 = z_0$), (2.17), and (2.18), we obtain in turn that

$$\begin{aligned}
\|x_1 - x^*\| &\leq \|z_0 - x^*\| \\
&+ \frac{\Big(3\varphi_0(\|x_0 - x^*\|) + \varphi_2(\|x_0 - x^*\|) + 2\Big)\int_0^1 \varphi_1\Big(\theta\|z_0 - x^*\|d\theta\|z_0 - x^*\|\Big)}{1 - \varphi_0(\|x_0 - x^*\|)} \\
&\leq \lambda_3(\|x_0 - x^*\|)\|x_0 - x^*\| \leq \|x_0 - x^*\|, \qquad (2.19)
\end{aligned}$$

which shows (2.10) for $n = 0$ and $x_1 \in U(x^*, R)$.

The induction for (2.8) - (2.10) is completed by replacing x_0, y_0, z_0, x_1 with x_m, y_m, z_m, x_{m+1}, respectively in the previous estimates. Then, by the estimate

$$\|x_{m+1} - x^*\| \leq c\|x_m - x^*\| < R, \tag{2.20}$$

where

$$c = \lambda_3(\|x_0 - x^*\|) \in [0, 1)$$

we can conclude that

$$\lim_{m \to +\infty} x_m = x^* \text{ and } x_{m+1} \in U(x^*, R).$$

Let $y^* \in D_1$ with $F(y^*) = 0$ and define linear operator

$$P = \int_0^1 F'\Big(x^* + \theta(y^* - x^*)\Big) d\theta.$$

Using (A_5), (A_1) and (A_2), we get in turn that

$$\begin{aligned} \|F'(x^*)^{-1}\Big(P - F'(x^*)\Big)\| &\leq \int_0^1 \varphi_0\Big(\theta\|x^* - y^*\|\Big) d\theta \\ &\leq \int_0^1 \varphi_0(\theta R^*) d\theta < 1, \end{aligned} \tag{2.21}$$

so $P^{-1} \in L(Y, X)$ and from the identity

$$0 = F(y^*) - F(x^*) = P(y^* - x^*), \tag{2.22}$$

we deduce that $x^* = y^*$. □

References

[1] Amat, S., Busquier, S. and Gutierrez, J.M., Geometric constructions of iterative functions to solve nonlinear equations, *J. Comput. Appl. Math.* 157, 2003, 197-205.

[2] Argyros, I.K., *Convergence and Applications of Newton-type Iterations*, Springer Verlag, Berlin, Germany, 2008.

[3] Argyros, I.K. and Chen, D., Results on the Chebsyhev method in Banach spaces, *Proyecciones* 12, 1993, 119-128.

[4] Argyros, I.K. and Magreñán, A.A., *A Contemporary Study of Iterative Methods*, Academic Press, Elsevier, San Diego, CA, USA, 2017.

[5] Argyros, I.K., Chen, D. and Qian, Q., A convergence analysis for rational methods with a parameter in Banach space, *Pure Math. Appl.* 5, 1994, 59-73.

[6] Argyros, I.K., On an improved unified convergence analysis for a certain class of Euler-Halley type methods, *J. Korean Soc. Math. Educ. Ser. B Pure Appl. Math.* 13, 2006, 207-215.

[7] Chen, D., Argyros, I.K. and Qian, Q., A local convergence theorem for the Supper-Halley method in a Banach space, *Appl. Math. Lett.* 7, 1994, 49-52.

[8] Ezquerro, J.A. and Hernandez, M.A., Halley's method for operators with unbounded second derivative, *Appl. Numer. Math.* 57, 2007, 354-360.

[9] Ezquerro, J.A. and Hernandez, M.A., New Kantorovich-type conditions for Halley's method, *Appl. Number. Anal. Comput. Math.* 2, 2005, 70-77.

[10] Ezquerro, J.A. and Hernandez, M.A., On the R-order of the Halley method, *J. Math. Anal. Appl.* 303, 2005, 591-601.

[11] Gutierrez, J.M. and Hernandez, M.A., A family of Chebyshev-Halley type methods in Banach space, *Bull. Aust. Math. Soc.* 55, 1997, 113-130.

[12] Gutierrez, J.M. and Hernandez, M.A., An acceleration of Newton's method: Super-Halley method, *Appl. Math. Comput.* 117, 2001, 223-239.

[13] Hernandez, M.A. and Salanova, M.A., A family of Chebyshev-Halley type methods. *Int. J. Comput. Math.* 47, 1993, 59-63.

[14] Huang, Z., On a family of Chebyshev-Halley type methods in Banach space under weaker Smale condition. *Number. Math. JCU* 9, 2000, 37-44.

[15] Kantorovich, L.V. and Akilov, G.P., *Functional Analysis in Normed Spaces*, Pergamon Press, New York, 1964.

Chapter 15

Convergence of a 2k Method

1. Introduction

The local convergence analysis of method

$$\begin{aligned}
w_1^{(n)} &= x^{(n)} - \frac{2}{3}u^{(n)},\\
w_2^{(n)} &= x^{(n)} - \left(I + \frac{21}{8}z^{(n)} - \frac{9}{2}\left(z_n^{(n)}\right)^2 + \frac{15}{8}\left(z^{(n)}\right)^3\right)u^{(n)},\\
w_3^{(n)} &= w_2^{(n)} - A_n V_3^{(n)},\\
w_4^{(n)} &= w_3^{(n)} - A_n V_4^{(n)},\\
&\vdots\\
x^{(n+1)} &= w_m = w_m^{(n)} - A_n V_m^{(n)},
\end{aligned} \tag{1.1}$$

where

$$\begin{aligned}
F'(x^{(n)})u^{(n)} &= F(x^{(n)}),\\
F'(x^{(n)})z^{(n)} &= F(y^{(n)}),\\
F'(x^{(n)})V_j^{(n)} &= F(w_{j-1}^{(n)}), j = 3, 4, \ldots, m,\\
y^{(0)} &= w_1^{(0)},
\end{aligned}$$

and $x_0 \in D$ was studied, when $X = Y = \mathbb{R}^k$. The convergence order was shown to be $2k$ under hypotheses up to the $2k+1$ derivative (although these derivatives do not appear in (1.1)) limiting the applicability of the method. As a motivational example, consider $D = \left[-\frac{3}{2}, \frac{1}{2}\right]$ and function F be defined on D by

$$F(x) = \begin{cases} x^3 \ln x^2 + x^5 - x^4, & x \neq 0 \\ 0, & x = 0. \end{cases}$$

Then, clearly $F'''(x)$ is unbounded on D. Hence, the earlier results cannot apply.

In this chapter, we show the local convergence analysis using only the first Fréchet-derivative, and in more general setting of Banach space valued operators. Error estimates and uniqueness results are also provide (not given in earlier works). The order of convergences uses computational order of convergence and approximate order of convergence not requiring high convergence order derivatives. This way, we expand the applicability of method and in the more general setting of a Banach space.

2. Local Convergence

Let $\varphi_0 : [0,\infty) \to [0,\infty)$ be a continuous and increasing function satisfying $\varphi_0(0) = 0$.
Suppose equation

$$\varphi_0(t) = 1 \tag{2.1}$$

has a minimal positive solution R_0. Let $\varphi : [0,R_0) \to [0,\infty)$ and $\varphi_1 : [0,R_0) \to [0,\infty)$ be continuous, and increasing functions with $\varphi(0) = 0$. Define functions λ_1 and μ_2 on the interval $[0,R_0)$ by

$$\lambda_1(t) = \frac{\int_0^1 \varphi((1-\theta)t)d\theta + \frac{1}{3}\int_0^1 \varphi_1(\theta t)d\theta}{1-\varphi_0(t)}$$

and

$$\mu_1(t) = \lambda_1(t) - 1.$$

Suppose that

$$\frac{\varphi_1(0)}{3} < 1. \tag{2.2}$$

We get by (2.2) that $\mu_1(0) = \frac{\varphi_1(0)}{3} - 1 < 0$, and by (2.1) $\mu_1(t) \to \infty$ as $t \to R_0^-$. By the intermediate value theorem equation $\mu_1(t) = 0$ has solution in $(0, R_0)$. Denote by R_1 the smallest such solution. Moreover, defined functions λ_2 and μ_2 on the interval $[0, R_0)$ by

$$\lambda_2(t) = \frac{\int_0^1 \varphi((1-\theta)t)d\theta}{1-\varphi_0(t)} + \frac{3\varphi_1(\lambda_1(t)t)\left[2+5\frac{\varphi_0(t)+\varphi(\lambda_1(t)t)}{1-\varphi_0(t)}\right]\left(\varphi_0(t)+\varphi(\lambda_1(t)t)\right)\int_0^1 \varphi_1(\theta t)d\theta}{8(1-\varphi_0(t))^3}$$

and

$$\mu_2(t) = \lambda_3(t) - 1.$$

We get $\mu_2(0) = -1$ and $\mu_2(t) \to \infty$ as $t \to R_0^-$. Denote by R_2 the smallest solution of equation $\mu_2(t) = 0$ in $(0, R_0)$. Furthermore, define functions $\lambda_j, \mu_j, j = 3, 4, \ldots, m$ on the interval $[0, R_0)$ by

$$\lambda_j(t) = \left(1 + \frac{a(t)}{1-\varphi_0(t)}\right)\lambda_{j-1}(t)$$

and

$$\mu_j(t) = \lambda_j(t) - 1,$$

where

$$a(t) = \frac{5(\varphi_0(t)+\varphi_0(\lambda_1(t)t))}{2(1-\varphi_0(t))} + \frac{1}{2}\frac{\varphi_1(\lambda_1(t)t)[2+\varphi_0(t)+\varphi_0(\lambda_1(t)t)]}{(1-\varphi_0(t))^2}$$

We obtain again that $\mu_j(0) = -1$ and $\mu_j \to \infty$ as $t \to R_0^-$. Denote by R_j the smallest solutions of equations $\mu_j(t) = 0$ in $(0, R_0)$, respectively:
Define a radius of convergence R by

$$R = \min\{R_j\}, j = 1, 2, 3, \ldots, m. \tag{2.3}$$

We have that for each $t \in [0,R)$

$$0 \le \varphi_(0) < 1 \tag{2.4}$$

$$0 \le a(t) \tag{2.5}$$

and

$$0 \le \lambda_j(t) < 1 \text{ for each } j = 1,2,3,\ldots,m. \tag{2.6}$$

Let $U(x,b), \overline{U}(x,b)$ stand for the open and closed balls in X, respectively with center $x \in X$ and of radius $b > 0$.
The local converges analysis of method (1.1) uses conditions (C):

(C_1) $F: D \subset X \to Y$ is a continuously Fréchet-differentiable operator, and there exists $x^* \in D$ such that

$$F(x^*) = 0 \text{ and } F'(x^*)^{-1} \in L(Y,X).$$

(C_2) There exists a function $\varphi_0 : [0,\infty) \to [0,\infty)$ continuous, increasing satisfying $\varphi_0(0) = 0$ such that

$$\|F'(x^*)^{-1}\Big(F'(x) - F'(x^*)\Big)\| \le \varphi_0(\|x - x^*\|).$$

Set $D_0 = D \cap U(x^*, R_0)$, where R_0 is given in (2.1).

(C_3) There exist functions $\varphi : [0,R_0) \to [0,\infty), \varphi_1 : [0,R_0) \to [0,\infty)$ such that for each $x, y \in D_0$

$$\|F'(x^*)^{-1}\Big(F'(x) - F'(y)\Big)\| \le \varphi(\|x - y\|),$$

and

$$\|F'(x^*)^{-1}F'(x)\| \le \varphi_1(\|x - x^*\|).$$

(C_4) $\overline{U}(x^*, R) \subseteq D$, (2.1) and (2.2) hold.

(C_5) There exists $R^* \le R$ such that

$$\int_0^1 \varphi(\theta R^*) d\theta < 1.$$

Set $D_1 = D \cap \overline{U}(x^*, R^*)$.

Next, we provide the local convergence analysis of method (1.1) using the conditions (C) and the aforementioned terminology.

Theorem 2.1 Under the conditions (C), iteration $\{x_n\}$ generated by method (1.1) for $x^{(0)} \in U(x^*, R) - \{x^*\}$ to x^*, which is the only solutions of equation $F(x) = 0$ in D_1 and

$$\|w_j^{(n)} - x^*\| \leq \lambda_j(\|w_j^{(n-1)} - x^*\|)\|w_j^{(n-1)} - x^*\| \leq \|w_j^{(n-1)} - x^*\| < R \quad (2.7)$$

for each $j = 1, 2, 3, \ldots, m$, where function λ_j, set D_1 are defined previously, and R is given in (2.3).

Proof. Let $x \in U(x^*, R)$. Then, by (2.1), (C_1), (C_2), and (2.4), we have in turn

$$\|F'(x^*)^{-1}\Big(F'(x) - F'(x^*)\Big)\| \leq \varphi_0(\|x - x^*\|) \leq \varphi_0(R < 1, \quad (2.8)$$

so by the Banach lemma on invertible operators [15], $F'(x)^{-1} \in L(Y, X)$, and

$$\|F'(x)^{-1}F'(x^*)\| \leq \frac{1}{1 - \varphi_0(\|x - x^*\|)}. \quad (2.9)$$

We also get that iterates $w_j^{(0)}$ are well defined by method (1.1) for $n = 0$, since by hypothesis $x^{(0)} \in U(x^*, R) - \{x^*\}$ and (2.9) holds for $x = x^{(0)}$. We can write by (C_1) that

$$F(x) = F(x) - F(x^*) = \int_0^1 F'\big(x^* + \theta(x - x^*)\big)d\theta(x - x^*). \quad (2.10)$$

Notice that $x^* + \theta(x - x^*) \in U(x^*, R)$ for each $\theta \in [0, 1]$, since

$$\|x^* + \theta(x - x^*) - x^*\| = \theta\|x - x^*\| < R.$$

Then, by the second hypothesis in (C_3) and (2.10), we get that

$$\|F'(x^*)^{-1}F(x)\| \leq \int_0^1 \varphi_1(\theta\|x - x^*\|)d\theta\|x - x^*\|. \quad (2.11)$$

In particular (2.11) holds for $x = x^{(0)}$. By the first sub-step of method (1.1) for $n = 1, j = 1$, we can write

$$w_1(0) - x^* = x^{(0)} - x^* - F'(x^{(0)})^{-1}F(x^{(0)}) + \frac{1}{3}F'(x^{(0)})^{-1}F(x^{(0)}) \quad (2.12)$$

In view of (2.3), (2.6)(for $j = 1$), the first and second hypothesis in (C_3), (2.9)(for $x = x^{(0)}$), (2.11)(for $x = x^{(0)}$), and (2.12), we have in turn that

$$\begin{aligned}
\|w_1^{(0)} - x^*\| &\leq \|F'(x^{(0)})^{-1}F'(x^*)\| \times \\
&\| \int_0^1 F'(x^{(0)})^{-1}\Big(F'(x^* + \theta(x^{(0)}) - F'(x^{(0)}))\Big)d\theta(x^{(0)} - x^*)\| \\
&+ \frac{1}{3}\|F'(x^{(0)})^{-1}F'(x^*)\|\|F'(x^*)^{-1}F(x^{(0)})\| \\
&\leq \frac{\int_0^1 \varphi((1-\theta)\|x^{(0)} - x^*\|)d\theta\|x^{(0)} - x^*\| + \frac{1}{3}\int_0^1 \varphi_1(\theta\|x^{(0)} - x^*\|)d\theta\|x^{(0)} - x^*\|}{1 - \varphi_0(\|x^{(0)} - x^*\|)} \\
&\leq \lambda_1(\|x^{(0)} - x^*\|)\|x^{(0)} - x^*\| \leq \|x^{(0)} - x^*\| < R, \quad (2.13)
\end{aligned}$$

which shows (2.7) for $n = 1, j = 1$ and $w_1(0) \in U(x^*, R)$.
We can write by the second sub-step of method (1.1) for $j = 2, n = 0$ that

$$\begin{aligned}
w_2^{(0)} - x^* &= x^{(0)} - x^* - F'(x^{(0)})^{-1}F(x^{(0)}) \\
&\quad - \frac{3}{8}F'(x^{(0)})F'(y^{(0)})\Big[7I - 12F'(x^{(0)})^{-1}F'(y^{(0)}) \\
&\quad + 5\Big(F'(x^{(0)})^{-1}F'(y^{(0)})\Big)^2\Big]F'(x^{(0)})^{-1}F(x^{(0)}) \\
&= x^{(0)} - x^* - F'(x^{(0)})^{-1}F(x^{(0)}) \\
&\quad - \frac{3}{8}F'(x^{(0)})^{-1}F'(y^{(0)})\Big[2I + 5F'(x^{(0)})^{-1}(F'(x^{(0)}) - F'(y^{(0)}))\times \\
&\quad F'(x^{(0)})^{-1}\Big(F'(x^{(0)}) - F'(y^{(0)})\Big)\Big]F'(x^{(0)})^{-1}F(x^{(0)}). \quad (2.14)
\end{aligned}$$

Using (2.3), (2.6)(for $i = 2$), (2.9), (2.11) - (2.14), we have in turn that

$$\|w_2^{(0)} - x^*\| \leq \frac{\int_0^1 \varphi((1-\theta)\|x^{(0)} - x^*\|)d\theta\|x^{(0)} - x^*\|}{1 - \varphi_0(\|x^{(0)} - x^*\|)}$$
$$+ \frac{3}{8}\frac{\varphi_1(b) + \left[2 + \frac{5\left(\varphi_0(b)+\varphi_0(\lambda_1(b)b)\right)\left(\varphi_0(b)+\varphi_0(\lambda_1(b)b)\right)\int_0^1 \varphi_1(\theta b)d\theta b}{1-\varphi_0(b)}\right]}{1-\varphi_0(b)}$$
$$= \lambda_2(b)b \leq b, \tag{2.15}$$

where,

$$b = \|x^{(0)} - x^*\|,$$

which shows (2.7) for $n = 0, j = 2$ and $w_2^{(0)} \in U(x^*, R)$.
We need an estimate on A_0 :

$$\begin{aligned}
A_0 &= 3I - \frac{5}{2}F'(x^{(0)})^{-1}F'(x^{(0)}) + \frac{1}{2}\left(F'(x^{(0)})^{-1}F'(y^{(0)})\right)^2 \\
&= \frac{5}{2}F'(x^{(0)})^{-1}\left(F'(x^0) - F'(y^0)\right) \\
&\quad + \frac{1}{2}F'(x^{(0)})^{-1}F'(y^{(0)})\left(I + F'(x^{(0)})^{-1}F'(y^{(0)})\right) \\
&= \frac{5}{2}F'(x^{(0)})^{-1}\left(F'(x^{(0)}) - F'(y^{(0)})\right) + \frac{1}{2}F'(x^{(0)})^{-1}F'(y^{(0)})F'(x^{(0)})^{-1}\times \\
&\quad \left[(F'(x^{(0)}) - F'(x^*)) + (F'(y^{(0)}) - F'(x^*)) + 2F'(x^*)\right].
\end{aligned} \tag{2.16}$$

Then, using the definition of function a, (2.9), (C_2), (C_3), (2.13), (2.15), and (2.16), we get in turn that

$$\begin{aligned}
\|A_0\| &\leq \frac{5\left(\varphi_0(\|x^{(0)} - x^*\|) + \varphi_0(\|y^{(0)} - x^*\|)\right)}{2\left(1 - \varphi_0(\|x^{(0)} - x^*\|)\right)} \\
&\quad + \frac{\varphi_1(\|y^{(0)} - x^*\|)\left[2 + \varphi_0(\|x^{(0)} - x^*\|) + \varphi_0(\|y^{(0)} - x^*\|)\right]}{2(1 - \varphi_0(\|x^{(0)} - x^*\|))^2} \\
&\leq a(\|x_0 - x^*\|).
\end{aligned} \tag{2.17}$$

In view of the third sub-step of method (1.1) for $j = 3, n = 0$, (2.3), (2.5), (2.6)(for $j = 3$), (2.15) and (2.17), we obtain in turn that

$$\begin{aligned}\|w_3^{(0)} - x^*\| &\leq \|w_2^{(0)} - x^*\| + \|A_0\|\|V_3^{(0)}\| \\ &\leq \Big(1 + \frac{a(\|x^{(0)} - x^*\|)}{1 - \varphi_0(\|x^{(0)} - x^*\|)}\Big)\|w_2^{(0)} - x^*\| \\ &\leq \lambda_3(\|x^{(0)} - x^*\|)\|x^{(0)} - x^*\| \leq \|x^{(0)} - x^*\|, \end{aligned} \tag{2.18}$$

which shows (2.7) for $j = 0, n = 0$ and $w_3^{(0)} \in U(x^*, R)$. Then, similarly for $j = 4, \ldots, m$

$$\begin{aligned}\|w_j^{(0)} - x^*\| &\leq \Big(1 + \frac{a(\|x^{(0)} - x^*\|)}{1 - \varphi_0(\|x^{(0)} - x^*\|)}\Big)\|w_{j-1}^{(0)} - x^*\| \\ &\leq \lambda_j(\|x^{(0)} - x^*\|)\|x^{(0)} - x^*\| \leq \|x^{(0)} - x^*\| < R, \end{aligned} \tag{2.19}$$

which shows (2.7) for $n = 0, j = 4, \ldots, m$ and $w_j^{(0)} \in U(x^*, r)$. Notice that $x^{(1)} = w_m$, so from (2.19), we can write that

$$\|x^1 - x^*\| \leq \lambda_m(\|x^{(0)} - x^*\|)\|x^{(0)} - x^*\|. \tag{2.20}$$

The induction for (2.7) is completed by simply replacing $w_0^{(0)}, w_1^{(0)}, \ldots, w_m^{(0)}$ by $w_0^{(k)}, w_1^{(k)}, \ldots, w_m^{(k)}$ in the preceding estimates. Then, using the estimate

$$\|x^{(k+1)} - x^*\| \leq c\|x^{(k)} - x^*\| < R, \tag{2.21}$$

we conclude that

$$\lim_{k \to +\infty} x^{(k)} = x^* \text{ and } x^{(k+1)} \in U(x^*, R).$$

It remains to show the uniqueness of the solution $x^* \in D_1$. Define

$$Q = \int_0^1 F'\Big(x^* + \theta(y^* - x^*)\Big)d\theta$$

for $y^* \in D_1$ with $F(y^*) = 0$. Using (C_2) and (C_5), we can get

$$\|F'(x^*)^{-1}\Big(Q - F'(x^*)\Big)\| \leq \int_0^1 \varphi_0(\theta\|x^* - y^*\|)d\theta \leq \int_0^1 \varphi_0(\theta R^*)d\theta < 1, \tag{2.22}$$

so $Q^{-1} \in L(Y,X)$. Finally, from the identity

$$0 = F(y^*) - F(x^*) = Q(y^* - x^*), \tag{2.23}$$

we conclude that $x^* = y^*$.

□

References

[1] Amat, S., Busquier, S. and Gutierrez, J.M., Geometric constructions of iterative functions to solve nonlinear equations, *J. Comput. Appl. Math.* 157, 2003, 197-205.

[2] Argyros, I.K., *Convergence and Applications of Newton-type Iterations*, Springer Verlag, Berlin, Germany, 2008.

[3] Argyros, I.K. and Chen, D., Results on the Chebsyhev method in Banach spaces, *Proyecciones* 12, 1993, 119-128.

[4] Argyros, I.K. and Magreñán, A.A., *A Contemporary Study of Iterative Methods*, Academic Press, Elsevier, San Diego, CA, USA, 2017.

[5] Argyros, I.K., Chen, D. and Qian, Q., A convergence analysis for rational methods with a parameter in Banach space, *Pure Math. Appl.* 5, 1994, 59-73.

[6] Argyros, I.K., On an improved unified convergence analysis for a certain class of Euler-Halley type methods, *J. Korean Soc. Math. Educ. Ser. B Pure Appl. Math.* 13, 2006, 207-215.

[7] Chen, D., Argyros, I.K. and Qian, Q., A local convergence theorem for the Supper-Halley method in a Banach space, *Appl. Math. Lett.* 7, 1994, 49-52.

[8] Ezquerro, J.A. and Hernandez, M.A., Halley's method for operators with unbounded second derivative, *Appl. Numer. Math.* 57, 2007, 354-360.

[9] Ezquerro, J.A. and Hernandez, M.A., New Kantorovich-type conditions for Halley's method, *Appl. Number. Anal. Comput. Math.* 2, 2005, 70-77.

[10] Ezquerro, J.A. and Hernandez, M.A., On the R-order of the Halley method, *J. Math. Anal. Appl.* 303, 2005, 591-601.

[11] Gutierrez, J.M. and Hernandez, M.A., A family of Chebyshev-Halley type methods in Banach space, *Bull. Aust. Math. Soc.* 55, 1997, 113-130.

[12] Gutierrez, J.M. and Hernandez, M.A., An acceleration of Newton's method: Super-Halley method, *Appl. Math. Comput.* 117, 2001, 223-239.

[13] Hernandez, M.A. and Salanova, M.A., A family of Chebyshev-Halley type methods. *Int. J. Comput. Math.* 47, 1993, 59-63.

[14] Huang, Z., On a family of Chebyshev-Halley type methods in Banach space under weaker Smale condition. *Number. Math. JCU* 9, 2000, 37-44.

[15] Kantorovich, L.V. and Akilov, G.P., *Functional Analysis in Normed Spaces*, Pergamon Press, New York, 1964.

Chapter 16

Convergence of an Eighth Order Method

1. Introduction

The local convergence analysis of method

$$
\begin{aligned}
y_n &= x_n - \frac{1}{2}F'(x_n)^{-1}F(x_n) \\
z_n &= y_n - 5F'(x_n)^{-1}F(y_n) \\
w_n &= z_n - \frac{1}{5}F'(x_n)^{-1}\Big(-16F(y_n)+F(z_n)\Big) \\
x_{n+1} &= w_n - G(v_n)F'(x_n)^{-1}F(w_n)
\end{aligned}
\tag{1.1}
$$

where

$$v_n = I - 5F'(x_n)^{-1}[y_n, z_n; F],$$

and

$$[.,.;F] : D \times D \to L(X,Y)$$

is a divided difference of order one, and $G : X \to L(X,Y)$ is a continuous operator. $x_0 \in D$ was studied, when $X = Y = \mathbb{R}^k$. The convergence order was shown to be eight under hypotheses up to the ninth derivative (although these derivatives do not appear in (1.1)) limiting the applicability of the method. As a motivational example, consider $D = \left[-\frac{3}{2}, \frac{1}{2}\right]$ and function F be defined on D by

$$F(x) = \begin{cases} x^3 \ln x^2 + x^5 - x^4, & x \neq 0 \\ 0, & x = 0. \end{cases}$$

Then, clearly $F'''(x)$ is unbounded on D. Hence, the earlier results cannot apply.

In this chapter, we show the local convergence analysis using only the first Fréchet-derivative, and in more general setting of Banach space valued operators. Error estimates and uniqueness results are also provide (not given in earlier works). The order of convergences uses computational order of convergence and approximate order of convergence not requiring high convergence order derivatives. This way, we expand the applicability of method and in the more general setting of a Banach space.

2. Local Convergence

Let $w_0 : [0,\infty) \to [0,\infty)$ be a continuous and increasing function satisfying $w_0(0) = 0$.
Suppose equation

$$w_0(t) = 1 \tag{2.1}$$

has a minimal positive solution ρ_0. Let w, w_1, g be continuous and increasing functions with $w(0) = 0$. Define λ_i, μ_i for $i = 1,2,3,4$ on the interval $[0, \rho_0)$ by

$$\lambda_1(t) = \frac{\int_0^1 W\Big((1-\theta)t\Big)d\theta}{1 - w_0(t)},$$

$$\lambda_2(t) = \Big(1 + \frac{5\int_0^1 w_1(\theta t)d\theta}{1 - w_0(t)}\Big)\lambda_1(t),$$

$$\lambda_3(t) = \Big(1 + \frac{16\int_0^1 w_1(\theta\lambda_1(t)t)d\theta\lambda_1(t) + \int_0^1 w_1(\theta\lambda_2(t)t)d\theta\lambda_2(t)}{5(1 - w_0(t))}\Big)\lambda_2(t)t,$$

$$\lambda_4(t) = \Big(1 + \frac{g(t)\int_0^1 w_1(\theta\lambda_3(t)t)d\theta}{1-\varphi_0(t)}\Big)\lambda_3(t),$$

and

$$\mu_i(t) = \lambda_i(t) - 1.$$

We have that $\mu_i(0) = -1$ and $\mu_i(t) \longrightarrow \infty$ as $t \longrightarrow \rho_0^-$. It follows by the intermediate value theorem that equations $\mu_i(t) = 0$ have solutions in $(0,\rho_0)$. Denote by ρ_i be the smallest such solution of equation $\mu_i(t) = 0$.
Define a radius of convergence ρ by

$$\rho = \min\{\rho_i\}. \tag{2.2}$$

It then follows that for each $t \in [0,\rho)$

$$0 \le w_0(t) < 1 \tag{2.3}$$

and

$$0 \le \lambda_i(t) < 1. \tag{2.4}$$

Let $U(x,a), \overline{U}(x,a)$ stand, respectively the open and closed balls in X with center $x \in X$ and of radius $a > 0$. Let also $L(X,Y)$ stand for the space of bounded linear operators from X into Y.
The following conditions (H) are used in the local convergence analysis of method (1.1):

(H_1) $F : D \subset X \longrightarrow Y$ is a Fréchet-differentiable operator, and $[.,.;F] : D \times D \longrightarrow L(X,Y)$ a divided difference of order one.

(H_2) There exists $x^* \in D$ such that $F(x^*) = 0$ and $F'(x^*)^{-1} \in L(Y,X)$.

(H_3) There exists a continuous, increasing function $w_0 : [0,\infty) \longrightarrow [0,\infty)$ satisfying $w_0(0) = 0$ such that for each $x \in D$

$$\|F'(x^*)^{-1}\Big(F'(x) - F'(x^*)\Big)\| \le w_0(\|x - x^*\|).$$

Set $D_0 = D \cap U(x^*,\rho_0)$.

(H_4) There exist continuous, increasing functions $W : [0,\rho_0) \to [0,\infty), w_1 : [0,\rho_0) \to [0,\infty)$ and $g : [0,\rho_0) \to [0,\infty)$ satisfying $W(0) = 0$ such that for each $x, y \in D_0$ and a continuous operator $G : X \to L(X,Y)$

$$\|F'(x^*)^{-1}\Big(F'(x) - F'(y)\Big)\| \leq W(\|x-y\|),$$

$$\|F'(x^*)^{-1}F'(x)\| \leq w_1(\|x-x^*\|),$$

and

$$\|G(v)\| \leq g(\|x-x^*\|),$$

where

$$v = I - 5F'(x)^{-1}[y,z;F],$$

$$y = x - F'(x)^{-1}F(x), z = y - 5F'(x)^{-1}F(y).$$

(H_5) $\overline{U}(x^*,\rho) \subseteq D$, and ρ_0 is given by (2.1).

(H_6) There exist $\rho^* \geq \rho$ such that

$$\int_0^1 w_0(\theta\rho^*)d\theta < 1.$$

Set $D_1 = D \cap \overline{U}(x^*,\rho^*)$.

Next, the local convergence analysis of method (1.1) follows under the conditions (H) and the aforementioned terminology.

Theorem 2.1 Under the hypotheses (H) iteration $\{x_n\}$ converges to x^* provided that $x_0 \in U(x^*,\rho) - \{x^*\}$ so that x^* is the only solution of equation $F(x) = 0$ in D_1 and

$$\|y_n - x^*\| \leq \lambda_1(\|x_n - x^*\|)\|x_n - x^*\| \leq \|x_n - x^*\| < \rho, \tag{2.5}$$

$$\|z_n - x^*\| \leq \lambda_2(\|x_n - x^*\|)\|x_n - x^*\| \leq \|x_n - x^*\|, \tag{2.6}$$

$$\|w_n - x^*\| \leq \lambda_3(\|x_n - x^*\|)\|x_n - x^*\| \leq \|x_n - x^*\|, \tag{2.7}$$

and

$$\|x_{n+1} - x^*\| \leq \lambda_4(\|x_n - x^*\|)\|x_n - x^*\| \leq \|x_n - x^*\|, \tag{2.8}$$

where set D_1, functions λ_i were given previously, and the radius of convergence ρ is given in (2.2).

Proof. We show that iteration $\{x_n\}$ is well defined in $U(x^*,\rho)$, remains in $U(x^*,\rho)$ for each $n=0,1,2,\ldots$, and converges so that error bounds (2.5) - (2.8) hold. By (2.1), (2.4), (2.5), (H_1) and (H_3), we obtain in turn that

$$\|F'(x^*)^{-1}\Big(F'(x_0)-F'(x^*)\Big)\| \leq w_0(\|x_0-x^*\|) \leq w_0(\rho) < 1, \tag{2.9}$$

so $F'(x_0)^{-1} \in L(Y,X)$ by the Banach lemma on invertible operators [15], and

$$\|F'(x_0)^{-1}F'(x^*)\| \leq \frac{1}{1-w_0(\|x_0-x^*\|)}. \tag{2.10}$$

It also follows that y_0, z_0, w_0 and x_1 are well defined by method (1.1) for $n=0$. We can write

$$\begin{aligned} y_0-x^* &= x_0-x^*-F'(x_0)^{-1}F(x_0)\\ &= F'(x_0)^{-1}\Big[\int_0^1 \Big(F'(x^*+\theta(x_0-x^*))-F'(x_0)\Big)(x_0-x^*)d\theta\Big]. \end{aligned} \tag{2.11}$$

Then, by (2.4), (2.6) (for $i=1$), the first hypothesis in (H_4), (2.10) and (2.11), we have in turn that

$$\begin{aligned} \|y_0-x^*\| &\leq \|F'(x_0)^{-1}F'(x^*)\|\|\int_0^1 F'(x^*)^{-1}\Big(F'(x^*+\theta(x_0-x^*))-F'(x_0)\Big)d\theta(x_0-x^*)\|\\ &\leq \frac{\int_0^1 W\Big((1-\theta)\|x_0-x^*\|\Big)d\theta\|x_0-x^*\|}{1-w_0(\|x_0-x^*\|)}\\ &= \lambda_1(\|x_0-x^*\|)\|x_0-x^*\| \leq \|x_0-x^*\| < \rho, \end{aligned} \tag{2.12}$$

which shows (2.5) for $n=0$ and $y_0 \in U(x^*,\rho)$.
Using (H_2), we can write

$$F(x_0) = F(x_0)-F(x^*) = \int_0^1 F'\Big(x^*+\theta(x_0-x^*)\Big)d\theta\|x_0-x^*\|. \tag{2.13}$$

In view of the estimate $\|x^*+\theta(x_0-x^*)-x^*\| = \theta\|x_0-x^*\| < \rho$, we have that $x^*+\theta(x_0-x^*) \in U(x^*,\rho)$ for each $\theta \in [0,1]$. Then, by the second hypothesis in

(H_4), and (2.13), we have that

$$\|F'(x^*)^{-1}F(x_0)\| \leq \int_0^1 w_1(\theta\|x_0 - x^*\|)d\theta\|x_0 - x^*\|. \tag{2.14}$$

Using (2.4), (2.6)(for $i = 2$), (2.10), method (1.1), (2.12) and (2.14)(for $x_0 = y_0$), we have in turn that

$$\begin{aligned} \|z_0 - x^*\| &\leq \|y_0 - x^*\| + 5\frac{\int_0^1 w_1(\theta\|y_0 - x^*\|)d\theta\|y_0 - x^*\|}{1 - w_0(\|x_0 - x^*\|)} \\ &\leq \lambda_2(\|x_0 - x^*\|)\|x_0 - x^*\| \leq \|x_0 - x^*\|, \end{aligned} \tag{2.15}$$

which shows (2.6) for $n = 0$ and $z_0 \in U(x^*, \rho)$.
Similarly, we obtain from the third sub-step of method (1.1) and (2.6)(for $i = 3$), (2.4) that

$$\begin{aligned} \|w_0 - x^*\| &\leq \|z_0 - x^*\| \\ &+ \frac{1}{5}\frac{16\int_0^1 w_1(\theta\|y_0 - x^*\|)d\theta\|y_0 - x^*\| + \int_0^1 w_1(\theta\|z_0 - x^*\|)d\theta\|z_0 - x^*\|}{1 - w_0(\|x_0 - x^*\|)} \\ &\leq \lambda_3(\|x_0 - x^*\|)\|x_0 - x^*\| \leq \|x_0 - x^*\|, \end{aligned} \tag{2.16}$$

which shows (2.7) for $n = 0$ and $w_0 \in U(x^*, \rho)$. Next, by the last sub-step of method (1.1) (for $n = 0$), (2.4), (2.6) (for $i = 4$), the last hypothesis in (H_4), (2.10) and (2.16), we have in turn that

$$\begin{aligned} \|x_1 - x^*\| &\leq \|w_0 - x^*\| + \frac{g(\|x_0 - x^*\|)\int_0^1 w_1(\theta\|w_0 - x^*\|)d\theta\|w_0 - x^*\|}{1 - w_0(\|x_0 - x^*\|)} \\ &\leq \lambda_4(\|x_0 - x^*\|)\|x_0 - x^*\| \leq \|x_0 - x^*\|, \end{aligned} \tag{2.17}$$

which shows (2.8) for $n = 0$ and $x_1 \in U(x^*, \rho)$. By replacing x_0, y_0, z_0, w_0, x_1 by $x_m, y_m, z_m, w_m, x_{m+1}$ in the previous estimates, we finish the induction for (2.5) - (2.8). The estimate

$$\|x_{m+1} - x^*\| \leq d\|x_m - x^*\| < \rho, \tag{2.18}$$

leads to $\lim_{m\to+\infty} x_m = x^*$, and $x_{m+1} \in U(x^*, \rho)$. Let

$$Q = \int_0^1 F'\Big(x^* + \theta(y^* - x^*)\Big)d\theta$$

for some $y^* \in D_1$ with $F(y*) = 0$.
Using (H_6) and (H_3), we can have

$$\|F'(x^*)^{-1}\Big(Q - F'(x^*)\Big)\| \leq \int_0^1 w_0(\theta\|x^* - y^*\|)d\theta$$
$$\leq \int_0^1 w_0(\theta\rho^*)d\theta < 1, \qquad (2.19)$$

so $Q^{-1} \in L(Y, X)$. Finally, from the identity

$$0 = F(y^*) - F(x^*) = Q(y^* - x^*), \qquad (2.20)$$

we deduce that $x^* = y^*$.

□

References

[1] Amat, S., Busquier, S. and Gutierrez, J.M., Geometric constructions of iterative functions to solve nonlinear equations, *J. Comput. Appl. Math.* 157, 2003, 197-205.

[2] Argyros, I.K., *Convergence and Applications of Newton-type Iterations*, Springer Verlag, Berlin, Germany, 2008.

[3] Argyros, I.K. and Chen, D., Results on the Chebsyhev method in Banach spaces, *Proyecciones* 12, 1993, 119-128.

[4] Argyros, I.K. and Magreñán, A.A., *A Contemporary Study of Iterative Methods*, Academic Press, Elsevier, San Diego, CA, USA, 2017.

[5] Argyros, I.K., Chen, D. and Qian, Q., A convergence analysis for rational methods with a parameter in Banach space, *Pure Math. Appl.* 5, 1994, 59-73.

[6] Argyros, I.K., On an improved unified convergence analysis for a certain class of Euler-Halley type methods, *J. Korean Soc. Math. Educ. Ser. B Pure Appl. Math.* 13, 2006, 207-215.

[7] Chen, D., Argyros, I.K. and Qian, Q., A local convergence theorem for the Supper-Halley method in a Banach space, *Appl. Math. Lett.* 7, 1994, 49-52.

[8] Ezquerro, J.A. and Hernandez, M.A., Halley's method for operators with unbounded second derivative, *Appl. Numer. Math.* 57, 2007, 354-360.

[9] Ezquerro, J.A. and Hernandez, M.A., New Kantorovich-type conditions for Halley's method, *Appl. Number. Anal. Comput. Math.* 2, 2005, 70-77.

[10] Ezquerro, J.A. and Hernandez, M.A., On the R-order of the Halley method, *J. Math. Anal. Appl.* 303, 2005, 591-601.

[11] Gutierrez, J.M. and Hernandez, M.A., A family of Chebyshev-Halley type methods in Banach space, *Bull. Aust. Math. Soc.* 55, 1997, 113-130.

[12] Gutierrez, J.M. and Hernandez, M.A., An acceleration of Newton's method: Super-Halley method, *Appl. Math. Comput.* 117, 2001, 223-239.

[13] Hernandez, M.A. and Salanova, M.A., A family of Chebyshev-Halley type methods. *Int. J. Comput. Math.* 47, 1993, 59-63.

[14] Huang, Z., On a family of Chebyshev-Halley type methods in Banach space under weaker Smale condition. *Number. Math. JCU* 9, 2000, 37-44.

[15] Kantorovich, L.V. and Akilov, G.P., *Functional Analysis in Normed Spaces*, Pergamon Press, New York, 1964.

Chapter 17

Convergence for a Method Containing a Sum of Linear Operators

1. Introduction

The local convergence analysis of method

$$\begin{aligned} y_n &= \varphi(x_n), \\ x_{n+1} &= x_n - 2A_n^{-1}F(x_n), \\ A_n &= \sum_{i=1}^{m} \alpha_i F'(z_n^i), \\ z_n^i &= \frac{(1+\theta i)y_n + (1-\theta i)x_n}{2}, \end{aligned} \tag{1.1}$$

where $\theta_i \in \mathbb{R}$, $\varphi : X \to Y$ be a continuous iteration function, m a given positive integer, α_i be a linear operator on X, and $x_0 \in D$ was studied, when $X = Y = \mathbb{R}^k$. The convergence order was shown to be six under Conditions up to the seventh derivative (although these derivatives do not appear in (1.1)) limiting the applicability of the method. As a motivational example, consider $D = \left[-\frac{3}{2}, \frac{1}{2}\right]$

and function F be defined on D by

$$F(x) = \begin{cases} x^3 \ln x^2 + x^5 - x^4, & x \neq 0 \\ 0, & x = 0. \end{cases}$$

Then, clearly $F'''(x)$ is unbounded on D. Hence, the earlier results cannot apply.

In this chapter, we show the local convergence analysis using only the first Fréchet-derivative, and in more general setting of Banach space valued operators. Error estimates and uniqueness results are also provide (not given in earlier works). The order of convergences uses computational order of convergence and approximate order of convergence not requiring high convergence order derivatives. This way, we expand the applicability of method and in the more general setting of a Banach space.

2. Local Convergence

Let $w_0 : [0, \infty) \to [0, \infty)$ be a continuous and increasing function satisfying $w_0(0) = 0$. Suppose that equation

$$w_0(t) = 1 \tag{2.1}$$

has a minimal positive solution ρ_0. let $\Psi : [0, \rho_0) \to [0, \infty)$ be a continuous and increasing function with $\Psi(0) = 0$.
Moreover, Suppose that equation

$$\Psi(t) = 1 \tag{2.2}$$

has a minimal positive solutions $\rho_1 \in (0, \rho_0)$. Define function γ_0 on the interval $[0, \rho_1)$ by

$$\gamma(t) = b^{-1} \sum_{i=1}^{m} \|\alpha_i\| w_0\Big(\frac{((1+\theta i)\Psi(t) + (1-\theta i))t}{2}\Big)$$

where

$$b = \sum_{i=1}^{m} \|\alpha_i\| \text{ and } \sum_{i=1}^{m} \alpha_i \text{ invertible}$$

Suppose that equation

$$\gamma(t) = 1 \tag{2.3}$$

has a positive minimal solution $\rho_2 \in (0, \rho_1)$. Notice that $\gamma(0) = 0$.
Define functions δ, g, h on $[0, \rho_2)$ by

$$\delta(t) = \sum_0^1 \|\alpha_i\| w_0\Big(\frac{((1+\theta i)\Psi(t) + (1-\theta i))t}{2}\Big) + 2w_0(t) + \| - 2I + \sum_{i=1}^m \alpha_i\|,$$

$$g(t) = \frac{\int_0^1 W\Big((1-\theta)t\Big) d\theta}{1 - w_0(t)} + \frac{\delta(t)\int_0^1 w_1(\theta t) d\theta}{b\big(1 - w_0(t)\big)\big(1 - \gamma(t)\big)}$$

and

$$h(t) = g(t) - 1.$$

Suppose that

$$\delta(0) w_1(0) < 1. \tag{2.4}$$

We have that $h(0) = g(0) - 1 = \delta(0) w_1(0) - 1 < 0$ by (2.4) and $h(t) \to \infty$ as $t \to \rho_2^-$. Denote by ρ the smallest solution of equation $h(t) = 0$ which is guaranteed to exist by the intermediate value theorem. Then for each $t \in [0, \rho)$

$$0 \le w_0(t) < 1, \tag{2.5}$$

$$0 \le \Psi(t) < 1, \tag{2.6}$$

$$0 \le \gamma(t) < 1, \tag{2.7}$$

$$0 \le \delta(t), \tag{2.8}$$

and

$$0 \le g(t) < 1. \tag{2.9}$$

The set $U(x, d), \overline{U}(x, d)$ denote the open and closed balls in X, respectively with center $x \in X$ and of radius $d > 0$.
The local convergence analysis is based on the conditions (H):

(H_1) $F : D \subseteq X \to Y$ is a continuously Fréchet - differentiable operator and there exists $x^* \in D$ such that

$$F(x^*) = 0 \text{ and } F'(x^*)^{-1} \in L(Y, X).$$

(H_2) There exists continuous and increasing function $w_0 : [0,\infty) \longrightarrow [0,\infty)$ with $w_0(0) = 0$ such that for each $x \in D$

$$\|F'(x^*)^{-1}\Big(F'(x) - F'(x^*)\Big)\| \leq w_0(\|x - x^*\|)$$

Set $D_0 = D \cap U(x^*, \rho_0)$, where ρ_0 is given in (2.1).

(H_3) There exist continuous and increasing functions $W : [0,\rho_0) \longrightarrow [0,\infty), w_1 : [0,\rho_0) \longrightarrow [0,\infty)$ with $w_0(0) = 0$; continuous operator $\varphi : X \to Y$ and continuous, increasing function $\Psi : [0,\rho_0) \longrightarrow [0,\infty)$ with $\Psi(0) = 0$ such that for each $x, y \in D_0$.

$$\|F'(x^*)^{-1}\Big(F'(x) - F'(y)\Big)\| \leq W(\|x - y\|),$$

$$\|F'(x^*)^{-1}F'(x)\| \leq w_1(\|x - x^*\|),$$

and

$$\|\varphi(x) - x^*\| \leq \Psi(\|x - x^*\|)\|x - x^*\|.$$

(H_4) $\overline{U}(x^*, \rho) \subseteq D$, (2.1) - (2.4) hold, $b \neq 0$, and

$$\text{linear operator } \sum_{i=1}^{m} \alpha_i \text{ is invertible.}$$

(H_5) There exists $\rho^* \geq \rho$ such that

$$\int_0^1 w_0(\theta\rho^*)d\theta < 1.$$

Set $D_1 = D \cap U(x^*, \rho)$.

Next, we provide the local convergence analysis using the preceding terminology, and the Conditions (H).

Theorem 2.1 Suppose that the Conditions (H) hold. Then, iterates $\{x_n\}$ starting at $x_0 \in U(x^*, \rho)$ and generated by method (1.1) exists in $U(x^*, \rho)$, stays in $U(x^*,\rho)$ for each $n = 0,1,2,\ldots$, and converges to the unique solution of equation $F(x) = 0$ in D_1 so that

$$\|y_n - x^*\| \leq \Psi(\|x_n - x^*\|)\|x_n - x^*\| \leq \|x_n - x^*\| < \rho \tag{2.10}$$

and

$$\|x_{n+1} - x^*\| \leq g(\|x_n - x^*\|)\|x_n - x^*\| \leq \|x_n - x^*\|, \tag{2.11}$$

where ρ, D_1, Ψ and g are defined previously.

Proof. Estimates (2.10) and (2.11) are shown using mathematical induction. Using the hypothesis $x_0 \in U(x^*,\rho) - \{x^*\}$, the definition of ρ_0, ρ, (2.5) and (H_2) we obtain in turn that

$$\|F'(x^*)^{-1}\Big(F'(x_0) - F'(x^*)\Big)\| \leq w_0(\|x_0 - x^*\|) \leq w_0(\rho) < w_0(\rho_0) < 1, \tag{2.12}$$

so by the Banach lemma on invertible operators [15], $F'(x_0)^{-1} \in L(Y,X)$ and

$$\|F'(x_0)^{-1}F'(x^*)\| \leq \frac{1}{1 - w_0(\|x_0 - x^*\|)}. \tag{2.13}$$

The iterate y_0 is well defined by the first sub-step of method (1.1) for $n = 0$. We also get by (2.6) and the last hypothesis in (H_3) that

$$\|y_0 - x^*\| = \|\varphi(x_0) - x^*\| \leq \Psi(\|x_0 - x^*\|)\|x_0 - x^*\| \leq \|x_0 - x^*\| < \rho, \tag{2.14}$$

which shows (2.10) for $n = 0$ and $y_0 \in U(x^*,\rho)$. We must show tht $A_0^{-1} \in L(Y,X)$. Using the definition of δ, b, (2.7) and (2.14), we get in turn that

$$\begin{aligned}
&\|\Big(\sum_{i=1}^{m} \alpha_i F'(x^*)\Big)^{-1}(A_0 - F'(x^*))\| \\
&\leq b^{-1}\sum_{i=1}^{m} \|\alpha_i\| w_0(\|z_0^i - x^*\|) \\
&\leq b^{-1}\sum_{i=1}^{m} \|\alpha_i\| w_0\Big(\frac{((1+\theta i)\Psi(\|x_0 - x^*\|) + (1-\theta i))\|x_0 - x^*\|}{2}\Big) \\
&\leq \gamma(\|x_0 - x^*\|) \leq \gamma(\rho) < 1,
\end{aligned} \tag{2.15}$$

so $A_0^{-1} \in L(Y,X)$ and

$$\|A_0^{-1}F'(x^*)\| \leq \frac{1}{b(1-\gamma(t))}, \tag{2.16}$$

where, we also used

$$\begin{aligned}\|z_0^i - x^*\| &= \|\frac{(1+\theta i)y_0 + (1-\theta i)x_0}{2} - x^*\| \\ &\leq \frac{(1+\theta i)\|y_0 - x^*\| + (1-\theta i)\|x_0 - x^*\|}{2} \\ &\leq \frac{\Big((1+\theta i)\Psi(\|x_0 - x^*\|) + (1-\theta i)\Big)\|x_0 - x^*\|}{2} \\ &\leq \|x_0 - x^*\| < \rho \end{aligned} \tag{2.17}$$

(i.e $z_0^i \in U(x^*, \rho)$).
We also need an estimate on

$$\begin{aligned} &\|F'(x^*)^{-1}\Big(A - 2F'(x_0)\Big)\| \\ &= \|F'(x^*)^{-1}\Big(\sum_{i=1}^{m}\|\alpha_i\|\big(F'(z_0^i) - F'(x^*)\big)\Big) - 2(F'(x_0) - F'(x^*)) \\ &+ \Big(-2I + \sum_{i=1}^{m}\alpha_i\Big)F'(x^*)\| \\ &\leq \sum_{i=1}^{m}\|\alpha_i\|w_0\Big(\frac{((1+\theta i)\Psi(\|x_0 - x^*\|) + (1-\theta i))\|x_0 - x^*\|}{2}\Big) \\ &+ 2w_0(\|x_0 - x^*\|) + \| -2I + \sum_{i=1}^{m}\alpha_i\| \\ &= \delta(\|x_0 - x^*\|). \end{aligned} \tag{2.18}$$

Then, by the second sub-step of method (1.1), the definition of ρ, γ, δ, (2.8), (2.9), (2.13), (2.15), (2.16) and from (2.18),

$$x_1 - x^* = x_0 - x^* - F'(x_0)^{-1}F(x_0) + F'(x_0)^{-1}\Big(A_0 - 2F'(x_0)\Big)A_0^{-1}F(x_0), \tag{2.19}$$

so we have in turn that

$$
\begin{aligned}
&\|x_1 - x^*\| \leq \|x_0 - x^* - F'(x_0)^{-1}F(x_0)\| \\
&+ \|F'(x_0)^{-1}F'(x^*)\| \|F'(x^*)^{-1}\big(A_0 - 2F'(x_0)\big)\| A_0^{-1}F'(x^*)\| F'(x^*)^{-1}F(x_0)\| \\
&\leq \left[\frac{\int_0^1 W\Big((1-\theta)\|x_0 - x^*\|\Big)d\theta}{1 - w_0(\|x_0 - x^*\|)} \right. \\
&\left. + \frac{\delta(\|x_0 - x^*\|)\int_0^1 w_1(\theta\|x_0 - x^*\|)d\theta}{b\big(1 - w_0(\|x_0 - x^*\|)\big)\big(1 - \gamma(\|x_0 - x^*\|)\big)} \right] \|x_0 - x^*\| \\
&= g(\|x_0 - x^*\|)\|x_0 - x^*\| \leq \|x_0 - x^*\|, \qquad (2.20)
\end{aligned}
$$

which shows (2.11) for $n = 0$, and $x_1 \in U(x^*, \rho)$. The induction for (2.10) and (2.11) is finished by replacing x_0, y_0, z_0^i, x_1 by x_k, y_k, z_k^i, x_{k+1} in the aforementioned estimates. Then, by the estimate

$$\|x_{k+1} - x^*\| \leq q(\|x_k - x^*\|) \leq \rho, \qquad (2.21)$$

where $q = g(\|x_0 - x^*\|) \in [0, 1)$, we conclude that

$$\lim_{k \to +\infty} x_k = x^* \text{ and } x_{k+1} \in U(x^*, \rho).$$

Let

$$G = \int_0^1 F'\big(y^* + \theta(y^* - x^*)\big)d\theta$$

for some $y^* \in D_1$ with $F(y^*) = 0$. Using (H_2) and (H_5), we get that

$$\|F'(x^*)^{-1}\Big(G - F'(x^*)\Big)\| \leq \int_0^1 w_0(\theta\|y^* - x^*\|)d\theta \leq \int_0^1 w_0(\theta\rho^*)d\theta < 1, \qquad (2.22)$$

so $G^{-1} \in (Y, X)$. Finally, from the identity

$$0 = F(x^*) - F(y^*) = G(x^* - y^*).$$

we deduce that $x^* = y^*$.

□

References

[1] Amat, S., Busquier, S. and Gutierrez, J.M., Geometric constructions of iterative functions to solve nonlinear equations, *J. Comput. Appl. Math.* 157, 2003, 197-205.

[2] Argyros, I.K., *Convergence and Applications of Newton-type Iterations*, Springer Verlag, Berlin, Germany, 2008.

[3] Argyros, I.K. and Chen, D., Results on the Chebsyhev method in Banach spaces, *Proyecciones* 12, 1993, 119-128.

[4] Argyros, I.K. and Magreñán, A.A., *A Contemporary Study of Iterative Methods*, Academic Press, Elsevier, San Diego, CA, USA, 2017.

[5] Argyros, I.K., Chen, D. and Qian, Q., A convergence analysis for rational methods with a parameter in Banach space, *Pure Math. Appl.* 5, 1994, 59-73.

[6] Argyros, I.K., On an improved unified convergence analysis for a certain class of Euler-Halley type methods, *J. Korean Soc. Math. Educ. Ser. B Pure Appl. Math.* 13, 2006, 207-215.

[7] Chen, D., Argyros, I.K. and Qian, Q., A local convergence theorem for the Supper-Halley method in a Banach space, *Appl. Math. Lett.* 7, 1994, 49-52.

[8] Ezquerro, J.A. and Hernandez, M.A., Halley's method for operators with unbounded second derivative, *Appl. Numer. Math.* 57, 2007, 354-360.

[9] Ezquerro, J.A. and Hernandez, M.A., New Kantorovich-type conditions for Halley's method, *Appl. Number. Anal. Comput. Math.* 2, 2005, 70-77.

[10] Ezquerro, J.A. and Hernandez, M.A., On the R-order of the Halley method, *J. Math. Anal. Appl.* 303, 2005, 591-601.

[11] Gutierrez, J.M. and Hernandez, M.A., A family of Chebyshev-Halley type methods in Banach space, *Bull. Aust. Math. Soc.* 55, 1997, 113-130.

[12] Gutierrez, J.M. and Hernandez, M.A., An acceleration of Newton's method: Super-Halley method, *Appl. Math. Comput.* 117, 2001, 223-239.

[13] Hernandez, M.A. and Salanova, M.A., A family of Chebyshev-Halley type methods. *Int. J. Comput. Math.* 47, 1993, 59-63.

[14] Huang, Z., On a family of Chebyshev-Halley type methods in Banach space under weaker Smale condition. *Number. Math. JCU* 9, 2000, 37-44.

[15] Kantorovich, L.V. and Akilov, G.P., *Functional Analysis in Normed Spaces*, Pergamon Press, New York, 1964.

Chapter 18

Local Convergence of a Two-Step Chebyshev-Type Method

1. Introduction

The local convergence analysis of method

$$y_n = x_n - \left(I + \frac{1}{2}(I - \alpha L_n)^{-1} L_n\right) F'(x_n)^{-1} F(x_n)$$
$$x_{n+1} = y_n - [y_n, x_n; F]^{-1} F(y_n) \tag{1.1}$$

where

$$L_n = F'(x_n)^{-1} F''(x_n) F'(x_n)^{-1} F(x_n)$$

$\alpha \in \mathbb{R}$, $[.,.;F] : D \times D \to L(X,Y)$ is a divided difference of order one, and $x_0 \in D$ was studied, when $X = Y = \mathbb{R}^k$. The convergence order was shown to be six under hypotheses up to the seventh derivative (although these derivatives do not appear in (1.1)) limiting the applicability of the method. As a motivational example, consider $D = \left[-\frac{3}{2}, \frac{1}{2}\right]$ and function F be defined on D by

$$F(x) = \begin{cases} x^3 \ln x^2 + x^5 - x^4, & x \neq 0 \\ 0, & x = 0. \end{cases}$$

Then, clearly $F'''(x)$ is unbounded on D. Hence, the earlier results cannot apply.

In this chapter, we show the local convergence analysis using only the first Fréchet-derivative, and in more general setting of Banach space valued operators. Error estimates and uniqueness results are also provide (not given in earlier works). The order of convergences uses computational order of convergence and approximate order of convergence not requiring high convergence order derivatives. This way, we expand the applicability of method and in the more general setting of a Banach space.

2. Local Convergence

Let $I = [0, \infty)$. Let also $\varphi_0 : I \to I$ be a continuous and increasing function with $\varphi_0(0) = 0$.
Suppose that equation

$$\varphi_0(t) = 1 \tag{2.1}$$

has a minimal positive solution ρ_1. Set $I_1 = [0, \rho_1)$ and let $\varphi, \varphi_1, \varphi_2 : I_1 \to I$ be continuous, and increasing functions with $\varphi(0) = 0$. Define functions p, p_1 and q on I_1 by

$$p(t) = |\alpha| \frac{\varphi_2(t) \int_0^1 \varphi_1(\theta t) d\theta t}{1 - \varphi_0(t)},$$

$$p_1(t) = \frac{\varphi_2(t) \int_0^1 \varphi_1(\theta t) d\theta t}{1 - \varphi_0(t)},$$

and

$$q(t) = p(t) - 1.$$

We have $q(0) = -1$, and $q(t) \to \infty$ as $t \to \rho_1^-$. It follows that equation $q(t) = 0$ has solutions in $(0, \rho_1)$ by the intermediate value theorem. Denote by ρ_2 the smallest such zero. Set $I_2 = [0, \rho_2)$ and define functions Ψ_1 and $\overline{\Psi}_1$ on I_2 by

$$\Psi_1(t) = \frac{\int_0^1 \varphi((1-\theta)t) d\theta}{1 - \varphi_0(t)} + \frac{1}{2} \frac{p_1(t) \int_0^1 \varphi_1(\theta t) d\theta}{(1 - p(t))(1 - \varphi_0(t))},$$

and

$$\overline{\Psi}_1(t) = \Psi_1(t) - 1.$$

We get $\overline{\Psi}_1(0) = -1$ and $\overline{\Psi}_1(t) \to \infty$ s $t \to \rho_2^-$. Denote by r_1 the smallest solution of equation $\overline{\Psi}_1(t) = 0$ in $(0, \rho_2)$.
Suppose that equation

$$\varphi_3(t) = 1 \tag{2.2}$$

has a minimal positive solution ρ_3 in $(0, \rho_2)$, where $\varphi_3 : I_2 \to I$ is continuous and increasing function with $\varphi_3(0) = 0$. Set $I_3 = [0, \rho_3)$ and define functions Ψ_2 and $\overline{\Psi}_2$ on I_3 by

$$\Psi_2(t) = \Big(1 + \frac{\int_0^1 \varphi_1(\theta\Psi_1(t)t)d\theta}{1 - \varphi_3(t)}\Big)\Psi_1(t)$$

and

$$\overline{\Psi}_2(t) = \Psi_2(t) - 1.$$

We obtain $\overline{\Psi}_2(0) = -1$ and $\overline{\Psi}_2(t) \to \infty$ as $t \to \rho_4^-$. Denote by r_2 the smallest solution of equation $\overline{\Psi}_2(t) = 0$ in $(0, \rho_4)$. Furthermore, define a radius of convergence r by

$$r = \min\{r_i\} \; i = 1, 2. \tag{2.3}$$

It follows that

$$0 \leq \varphi_0(t) < 1, \tag{2.4}$$

$$0 \leq p(t) < 1, \tag{2.5}$$

$$0 \leq \Psi_1(t) < 1, \tag{2.6}$$

$$0 \leq \Psi_3(t) < 1, \tag{2.7}$$

and

$$0 \leq \Psi_2(t) < 1. \tag{2.8}$$

The sets $U(z, d), \overline{U}(z, d)$ denote, respectively the open and closed balls in X with center $z \in X$ and of radius $d > 0$.
The conditions (C) are involved in the local convergence analysis of method (1.1).

(C_1) $F : D \subset X \to Y$ is continuously Fréchet-differentiable operator, $[.,.;F] : D \times D \to L(X,Y)$ is a divided difference of order one and there exists $x^* \in D$ such that

$$F(x^*) = 0 \text{ and } F'(x^*)^{-1} \in L(Y,X).$$

(C_2) There exist function $\varphi_0 : I \to I$ continuous, increasing with $\varphi_0(0) = 0$ such that for each $x \in D$

$$\|F'(x^*)^{-1}\Big(F'(x) - F'(x^*)\Big)\| \leq \varphi_0(\|x - x^*\|).$$

Let $D_1 = D \cap U(x^*, \rho_1)$, where ρ_1 is given in (2.1).

(C_3) There exist functions $\varphi, \varphi_1, \varphi_2, \varphi_3 : I_1 \to I$ with $\varphi(0) = \varphi_3(0) = 0$ such that for each $x, y \in D_1$.

$$\|F'(x^*)^{-1}\Big(F'(x) - F'(y)\Big)\| \leq \varphi(\|x - y\|),$$
$$\|F'(x^*)^{-1}F'(x)\| \leq \varphi_1(\|x - x^*\|),$$
$$\|F'(x^*)^{-1}F''(x)\| \leq \varphi_2(\|x - x^*\|),$$

and

$$\|F'(x^*)^{-1}\Big([x,y;F] - F'(x^*)\Big)\| \leq \varphi_3(\|x - x^*\|),$$

where

$$y = x - \Big(I + \frac{1}{2}(I - \alpha L(x))^{-1}L(x)\Big)F'(x)^{-1}F(x).$$

(C_4) $\overline{U}(x^*, r) \subseteq X$, where r is given in (2.3).

(C_5) There exists $r^* \geq r$ such that

$$\int_0^1 \varphi_0(\theta r^*)d\theta < 1.$$

Let $D_1 = D \cap \overline{U}(x^*, r^*)$.

Theorem 2.1 Under the conditions (C), iteration $\{x_n\}$ starting at $x_0 \in U(x^*, r) - \{x^*\}$ converges to the only solution x^* of equation $F(x) = 0$ in D_1 so that

$$\|y_n - x^*\| \leq \Psi_1(\|x_n - x^*\|)\|x_n - x^*\| \leq \|x_n - x^*\| < r \tag{2.9}$$

and

$$\|x_{n+1} - x^*\| \leq \Psi_2(\|x_n - x^*\|)\|x_n - x^*\| \leq \|x_n - x^*\|, \tag{2.10}$$

where r, r^*, Ψ_1, Ψ_2 are defined previously.

Proof. Inequations (2.9) and (2.10) are proved using mathematical induction. It follows from $x_0 \in U(x^*, r) - \{x^*\}$, (2.1), (2.3), (2.4), and (C_2) that

$$\|F'(x^*)^{-1}\Big(F'(x_0) - F'(x^*)\Big)\| \leq \varphi_0(\|x_0 - x^*\|) \leq \varphi_0(r) < 1, \tag{2.11}$$

so $F'(x_0)^{-1} \in L(Y, X)$ and

$$\|F'(x_0)^{-1}F'(x^*)\| \leq \frac{1}{1 - \varphi_0(\|x_0 - x^*\|)}. \tag{2.12}$$

by the Banach lemma on invertible operators [15]. We can write

$$F(x_0) = F(x_0) - F(x^*) = \int_0^1 F'\big(x^* + \theta(x_0 - x^*)\big)d\theta(x_0 - x^*), \tag{2.13}$$

so by the second condition in (C_3) and (2.12), we have that

$$\|F'(x^*)^{-1}F(x_0)\| \leq \int_0^1 \varphi_1(\theta\|x_0 - x^*\|)d\theta\|x_0 - x^*\|, \tag{2.14}$$

where we also set $x^* + \theta(x_0 - x^*) \in U(x^*, r)$, since $\|x^* + \theta(x_0 - x^*)\| = \theta\|x_0 - x^*\| < r$. We get the estimates from (2.3), (2.5), (2.14), and second and third condition in (C_3) that

$$\begin{aligned}
\|\alpha L_0\| &\leq \|F'(x_0)^{-1}F'(x^*)\|\|F'(x^*)^{-1}F''(x_0)\| \times \\
&\qquad \|F'(x_0)^{-1}F'(x^*)\|\|F'(x^*)^{-1}F(x_0)\| \\
&\leq |\alpha|\frac{\varphi_2(\|x_0 - x^*\|)\int_0^1 \varphi_1(\theta\|x_0 - x^*\|)d\theta\|x_0 - x^*\|}{1 - \varphi_0(\|x_0 - x^*\|)} \\
&= p(\|x_0 - x^*\|) \leq p(r) < 1,
\end{aligned} \tag{2.15}$$

so $(I-\alpha L_0)^{-1} \in L(Y,X)$, and

$$\|(I-\alpha L_0)^{-1}\| \leq \frac{1}{1-p(\|x_0-x^*\|)}. \tag{2.16}$$

It also follows that y_0 is well defined by the the first sub-step of method (1.1) for $n=0$. Then, we can write

$$y_0-x^* = x_0-x^*-F'(x_0)^{-1}F(x_0)+\frac{1}{2}(I-\alpha L_0)^{-1}L_0F'(x_0)^{-1}F(x_0). \tag{2.17}$$

By (2.3), (2.6), first condition in (C_3), (2.12), (2.14), (2.16) and (2.17), first we have in turn that

$$\begin{aligned}
\|y_0-x^*\| &\leq \|x_0-x^*-F'(x_0)^{-1}F(x_0)\| \\
&\quad +\frac{1}{2}\|(I-\alpha L_0)^{-1}\|\|L_0\|\|F'(x_0)^{-1}F(x_0)\| \\
&\leq \|F'(x_0)^{-1}F'(x^*)\|\|\int_0^1 F'(x^*)^{-1}\Big(F'(x^*+\theta(x_0-x^*)) \\
&\qquad -F'(x_0)\Big)(x_0-x^*)d\theta\| \\
&\quad +\frac{1}{2}\|(I-\alpha L_0)^{-1}\|\|L_0\|\|F'(x_0)^{-1}F'(x^*)\|\|F'(x^*)^{-1}F(x_0)\| \\
&\leq \frac{\int_0^1 \varphi((1-\theta)\|x_0-x^*\|)d\theta\|x_0-x^*\|}{1-\varphi_0(\|x_0-x^*\|)} \\
&\quad +\frac{1}{2}\frac{p_1(\|x_0-x^*\|)\int_0^1 \varphi_1(\theta\|x_0-x^*\|)d\theta\|x_0-x^*\|}{\big(1-p(\|x_0-x^*\|)\big)\big(1-\varphi_0(\|x_0-x^*\|)\big)} \\
&\leq \Psi_1(\|x_0-x^*\|)\|x_0-x^*\| \leq \|x_0-x^*\| < r,
\end{aligned} \tag{2.18}$$

which shows (2.9) for $n=0$ and $y_0 \in U(x^*,r)$. We must show $[y_0,x_0;F]^{-1} \in L(Y,X)$. By the last conditions in (C_3), (2.2), and (2.18), we get

$$\|F'(x^*)^{-1}\Big([y_0,x_0;F]-F'(x^*)\Big)\| \leq \varphi_3(\|x_0-x^*\|) \leq \varphi_3(r) < 1, \tag{2.19}$$

so

$$\|[y_0,x_0;F]^{-1}F'(x^*)\| \leq \frac{1}{1-\varphi_3(\|x_0-x^*\|)}, \tag{2.20}$$

and x_1 is well defined by the last sub-step of method (1.1) for $n = 0$. Then, by (2.3), (2.8), (2.14)(for $x_0 = y_0$), and (2.20) that

$$\begin{aligned}\|x_1-x^*\| &\leq \|y_0-x^*\| + \|[y_0,x_0;F]^{-1}F'(x^*)\|\|F'(x^*)^{-1}F(y_0)\| \\ &\leq \left(1+\frac{\int_0^1 \varphi_1(\theta\|y_0-x^*\|)d\theta}{1-\varphi_3(\|x_0-x^*\|)}\right)\|y_0-x^*\| \\ &\leq \Psi_2(\|x_0-x^*\|)\|x_0-x^*\| \leq \|x_0-x^*\|, \end{aligned} \tag{2.21}$$

which shows (2.10) for $n = 0$ and $x_1 \in U(x^*,r)$. Next, replace x_0, y_0, x_1 by x_m, y_m, x_{m+1} in the previous estimates to finish the induction for (2.9) and (2.10). It then follows from the estimate

$$\|x_{m+1}-x^*\| \leq q\|x_m-x^*\| < r, \tag{2.22}$$

where $q = \Psi_2(\|x_0-x^*\|) \in [0,1)$ that

$$\lim_{m\longrightarrow+\infty} x_m = x^* \text{ and } x_{m+1} \in U(x^*,r).$$

Set

$$T = \int_0^1 F'(x^*+\theta(y^*-x^*))d\theta$$

for $y^* \in D_1$ with $F(y^*) = 0$. In view of (C_2) and (C_5), we get

$$\|F'(x^*)^{-1}\Big(T-F'(x)\Big)\| \leq \int_0^1 \varphi_0(\theta\|x^*-y^*\|)d\theta \leq \int_0^1 \varphi_0(\theta r^*)d\theta < 1, \tag{2.23}$$

so T is invertible. Finally, from the identity

$$0 = F(y^*) - F(x^*) = T(y^*-x^*), \tag{2.24}$$

we deduce that $x^* = y^*$.

□

References

[1] Amat, S., Busquier, S. and Gutierrez, J.M., Geometric constructions of iterative functions to solve nonlinear equations, *J. Comput. Appl. Math.* 157, 2003, 197-205.

[2] Argyros, I.K., *Convergence and Applications of Newton-type Iterations*, Springer Verlag, Berlin, Germany, 2008.

[3] Argyros, I.K. and Chen, D., Results on the Chebsyhev method in Banach spaces, *Proyecciones* 12, 1993, 119-128.

[4] Argyros, I.K. and Magreñán, A.A., *A Contemporary Study of Iterative Methods*, Academic Press, Elsevier, San Diego, CA, USA, 2017.

[5] Argyros, I.K., Chen, D. and Qian, Q., A convergence analysis for rational methods with a parameter in Banach space, *Pure Math. Appl.* 5, 1994, 59-73.

[6] Argyros, I.K., On an improved unified convergence analysis for a certain class of Euler-Halley type methods, *J. Korean Soc. Math. Educ. Ser. B Pure Appl. Math.* 13, 2006, 207-215.

[7] Chen, D., Argyros, I.K. and Qian, Q., A local convergence theorem for the Supper-Halley method in a Banach space, *Appl. Math. Lett.* 7, 1994, 49-52.

[8] Ezquerro, J.A. and Hernandez, M.A., Halley's method for operators with unbounded second derivative, *Appl. Numer. Math.* 57, 2007, 354-360.

[9] Ezquerro, J.A. and Hernandez, M.A., New Kantorovich-type conditions for Halley's method, *Appl. Number. Anal. Comput. Math.* 2, 2005, 70-77.

[10] Ezquerro, J.A. and Hernandez, M.A., On the R-order of the Halley method, *J. Math. Anal. Appl.* 303, 2005, 591-601.

[11] Gutierrez, J.M. and Hernandez, M.A., A family of Chebyshev-Halley type methods in Banach space, *Bull. Aust. Math. Soc.* 55, 1997, 113-130.

[12] Gutierrez, J.M. and Hernandez, M.A., An acceleration of Newton's method: Super-Halley method, *Appl. Math. Comput.* 117, 2001, 223-239.

[13] Hernandez, M.A. and Salanova, M.A., A family of Chebyshev-Halley type methods. *Int. J. Comput. Math.* 47, 1993, 59-63.

[14] Huang, Z., On a family of Chebyshev-Halley type methods in Banach space under weaker Smale condition. *Number. Math. JCU* 9, 2000, 37-44.

[15] Kantorovich, L.V. and Akilov, G.P., *Functional Analysis in Normed Spaces*, Pergamon Press, New York, 1964.

Chapter 19

Local Convergence of a Three Parameter Sixth Order Method

1. Introduction

The local convergence analysis of method

$$\begin{aligned} y_n &= x_n - \frac{3}{2}F'(x_n)^{-1}F(x_n), \\ z_n &= x_n - \left[I + \frac{3C_n}{4}A_n^{-1}C_n\right]F'(x_n)^{-1}F(x_n), \\ x_{n+1} &= z_n - B_n^{-1}\left(F'(x_n) + \delta F'(y_n)\right)F'(x_n)^{-1}F(z_n), \end{aligned} \tag{1.1}$$

where

$$\begin{aligned} A_n &= I + 6(4I - 3\alpha C_n), \\ C_n &= I - F'(x_n)^{-1}F'(y_n), \\ B_n &= \beta F'(x_n) + \gamma F'(y_n) \end{aligned}$$

$\gamma \in \mathbb{R} - \{-1\}, \beta = \frac{2-3\gamma}{5}, \delta = \frac{2\gamma-3}{5}, \alpha \in \mathbb{R}$, and $x_0 \in D$ was studied, when $X = Y = \mathbb{R}^k$. The convergence order was shown to be six under conditions up to the seventh derivative (although these derivatives do not appear in (1.1)) limiting the applicability of the method. As a motivational example, consider $D = \left[-\frac{3}{2}, \frac{1}{2}\right]$ and function F be defined on D by

$$F(x) = \begin{cases} x^3 \ln x^2 + x^5 - x^4, & x \neq 0 \\ 0, & x = 0. \end{cases}$$

Then, clearly $F'''(x)$ is unbounded on D. Hence, the earlier results cannot apply.

In this chapter, we show the local convergence analysis using only the first Fréchet-derivative, and in more general setting of Banach space valued operators. Error estimates and uniqueness results are also provide (not given in earlier works). The order of convergences uses computational order of convergence and approximate order of convergence not requiring high convergence order derivatives. This way, we expand the applicability of method and in the more general setting of a Banach space.

2. Local Convergence

Let $S = [0, \infty)$. Let also $\Omega_0 : I \to I$ be a continuous and increasing function with $\Omega_0(0) = 0$.
Suppose that equation

$$\Omega_0(t) = 1 \tag{2.1}$$

has a minimal positive solution ρ_0. Set $S_0 = [0, \rho_0)$, and let also functions $\Omega : S_0 \to S, \Omega_1 : S_0 \to S$ be continuous and increasing with $\Omega(0) = 0$. Define functions λ_1 and μ_1 on so by

$$\lambda_1(t) = \frac{\int_0^1 \Omega\Big((1-\theta)t\Big)d\theta + \frac{1}{3}\int_0^1 \Omega_1(\theta t)d\theta}{1 - \Omega_0(t)}$$

and

$$\mu_1(t) = \lambda_1(t) - 1.$$

Suppose that

$$\frac{\Omega_1(0)}{3} - 1 < 0. \tag{2.2}$$

We have $\mu_1(0) < 0$ by (2.2) and $\mu_1(t) \to \infty$ as $t \to \rho_0^-$. Denote by ρ_1 the smallest solution of equation $\mu_1(t) = 0$ in $(0, \rho_0)$ guaranteed to exist by the intermediate value theorem. Define function P on S_0 by

$$P(t) = \frac{\Omega_0(t) + \Omega(\lambda_1(t)t)}{1 - \Omega_0(t)}.$$

Define functions p_1 and q_1 on S_0 by

$$p_1(t) = 6\left[|4 - 3\alpha| + \frac{3|\alpha|\Omega_1(\lambda_1(t)t)}{1 - \Omega_0(t)}\right]$$

and

$$q_1(t) = p_1(t) - 1.$$

Suppose that

$$6[|4 - 3\alpha| + 6|\alpha|\Omega_1(0)] - 1 < 0. \tag{2.3}$$

We get $q_1(0) < 0$ and $q_1(t) \to \infty$ as $t \to \rho_0^-$. Denote by ρ_{q_1} the smallest solution of equation $q_1(t) = 0$ in $(0, \rho_0)$.
Define functions λ_2 and μ_2 on S_0 by

$$\lambda_2(t) = \frac{\int_0^1 \Omega((1-\theta)t)d\theta}{1 - \Omega_0(t)} + \frac{3P(t)^2 \int_0^1 \Omega_1(\theta t)d\theta}{4(1 - p_1(t))(1 - \Omega_0(t))}$$

and

$$\mu_2(t) = \lambda_2(t) - 1.$$

We have $\mu_2(0) = -1$ and $\mu_2(t) \to \infty$ as $t \to \rho_{q_1}^-$. Denote by r_2 the smallest such solution of equation $\mu_2(t) = 0$ in $(0, \rho_{q_1})$. Define functions p_2 and q_2 on S_0 by

$$p_2(t) = |\beta + \gamma|^{-1}\left[|\beta|\Omega_0(t) + |\gamma|\Omega_0(\lambda_1(t)t)\right]$$

and

$$q_2(t) = p_2(t) - 1.$$

We get $q_2(0) = -1$ and $q_2(t) \to \infty$ as $t \to \rho_0^-$. Denote by ρ_{q_2} the smallest solution of equation $q_2(t) = 0$ in $(0, \rho_0)$. Let $S_1 = [0, \rho_{q_2})$. Define functions λ_3

and μ_3 on S_1 by

$$\lambda_3(t) = \left(1 + \frac{\left(\Omega_1(t) + |\rho|\Omega_1(\lambda_1(t)t)\right)\int_0^1 \Omega_1\left(\theta\lambda_2(t)t\right)d\theta}{|\beta+\gamma|\left(1 - p_2(t)\right)\left(1 - \Omega_0(t)\right)}\right)\lambda_2(t)$$

and

$$\mu_3(t) = \lambda_3(t) - 1.$$

We obtain $\mu_3(0) = -1$ and $\mu_3(t) \to \infty$ as $t \to \rho_{q2}^-$. Denote by ρ_3 the smallest such solution of equation $\mu_3(t) = 0$ in $(0, \rho_{q_2})$.
Define a radius of convergence ρ by

$$\rho = \min\{\rho_i\}, i = 1,2,3. \tag{2.4}$$

It follows that for each $t \in [0,\rho)$.

$$0 \leq \Omega_0(t) < 1, \tag{2.5}$$

$$0 \leq \lambda_i(t) < 1, \tag{2.6}$$

$$0 \leq P(t) < 1, \tag{2.7}$$

and (2.8)

$$0 \leq p_2(t) < 1. \tag{2.9}$$

The sets $U(v,\varepsilon), \overline{U}(v,\varepsilon)$ stand respectively for the open and closed balls in X with center $x \in X$ and of radius $\varepsilon > 0$.
We shall base the local convergence analysis of method (1.1) on the conditions (H):

(H_1) $F : D \subset X \to Y$ is a continuously Fréchet-differentiable operator and there exists $x^* \in D$ such that $F(x^*) = 0$ and $F'(x^*)^{-1} \in L(Y,X)$.

(H_2) There exists a function $\Omega_0 : S \to S$ continuous and increasing with $\Omega_0(0) = 0$ such that for each $x \in D$.

$$\|F'(x^*)^{-1}\left(F'(x) - F'(x^*)\right)\| \leq \Omega_0(\|x - x^*\|).$$

Set $D_0 = D \cap U(x^*, \rho_0)$, where ρ_0 is given in (2.1).

(H_3) There exist functions $\Omega : S_0 \to S, \Omega_1 : S_0 \to S$ continuous and increasing function such that for each $x, y \in D_0$

$$\|F'(x^*)^{-1}\Big(F'(x) - F'(x^*)\Big)\| \leq \Omega(\|x - y\|)$$

and

$$\|F'(x^*)^{-1}F'(x)\| \leq \Omega_1(\|x - x^*\|).$$

(H_4) $\overline{U}(x^*, \rho) \subset D$, (2.2) and (2.3) hold.

(H_5) There exists $\rho^* \geq \rho$ such that

$$\int_0^1 \Omega_0(\theta\rho^*)d\theta < 1.$$

Set $D_1 = D \cap \overline{U}(x^*, \rho)$.

Theorem 2.1 Under the conditions (H), iteration $\{x_n\}$ produced by method (1.1) converges to the only solution x^* of equation $F(x) = 0$ in D_1 provided that $x_0 \in U(x^*, \rho) - \{x^*\}$,

$$\|y_n - x^*\| \leq \lambda_1(\|x_n - x^*\|)\|x_n - x^*\| \leq \|x_n - x^*\| < \rho, \tag{2.10}$$

$$\|z_n - x^*\| \leq \lambda_2(\|x_n - x^*\|)\|x_n - x^*\| \leq \|x_n - x^*\|, \tag{2.11}$$

and

$$\|x_{n+1} - x^*\| \leq \lambda_3(\|x_n - x^*\|)\|x_n - x^*\| \leq \|x_n - x^*\|. \tag{2.12}$$

where ρ is given in (2.4).

Proof. Inequations (2.10) - (2.12) are proved using mathematical induction. By $x_0 \in U(x^*, \rho) - \{x^*\}$, (2.4), (2.5), and (H_2), we obtain in turn that

$$\|F'(x^*)^{-1}\Big(F'(x_0) - F'(x^*)\Big)\| \leq \Omega_0(\|x_0 - x^*\|) \leq \Omega_0(\rho) < 1, \tag{2.13}$$

so by the Banach lemma on invertible operators [15], $F'(x_0)^{-1} \in L(Y, X)$, and

$$\|F'(x_0)^{-1}F'(x^*)\| \leq \frac{1}{1 - \Omega_0(\|x_0 - x^*\|)} \tag{2.14}$$

It also follows that y_0 is well defined the first sub-step of method (1.1) for $n = 0$. We can write by (H_1)

$$F(x_0) = F(x_0) - F(x^*) = \int_0^1 F'\Big(x^* + \theta(x_0 - x^*)\Big)d\theta(x_0 - x^*). \tag{2.15}$$

Then, by the second conditions in (H_2) and (2.15), we get

$$\|F'(x^*)^{-1}F(x_0)\| \le \int_0^1 \Omega_1(\theta\|x_0 - x^*\|)d\theta\|x_0 - x^*\|, \tag{2.16}$$

where we also used $x^* + \theta(y^* - x^*) \in \overline{U}(x^*, \rho)$ for each $\theta \in [0, 1]$ since

$$\|x^* + \theta(y^* - x^*) - x^*\| = \theta\|y^* - x^*\| < \rho.$$

Using the first sub-step of method (1.1), (2.4), (2.6) (for $i = 1$), (2.14), (2.16) and the first hypothesis in (H_2) that

$$\begin{aligned}
\|y_0 - x^*\| &= \|x_0 - x^* - F'(x_0)^{-1}F(x_0) + \frac{1}{2}F'(x_0)^{-1}F(x_0)\| \\
&\le \|F'(x_0)^{-1}F'(x^*)\|\|\int_0^1 F'(x^*)^{-1}\Big(F'(x^* + \theta(x_0 - x^*)) - F'(x_0)\Big)(x_0 - x^*)d\theta\| \\
&+ \frac{1}{3}\|F'(x_0)^{-1}F'(x^*)\|\|F'(x^*)^{-1}F(x_0)\| \\
&\le \frac{\int_0^1 \Omega\Big((1-\theta)\|x_0 - x^*\|d\theta\Big) + \frac{1}{3}\int_0^1 \Omega_1(\theta\|x_0 - x^*\|)d\theta}{1 - \Omega_0(\|x_0 - x^*\|)}\|x_0 - x^*\| \\
&\le \lambda_1(\|x_0 - x^*\|) \le \|x_0 - x^*\| < \rho,
\end{aligned} \tag{2.17}$$

which shows (2.10) for $n = 0$, and $y_0 \in \overline{U}(x^*, \rho)$.
We need an estimate on $\|C_0\|$:

$$\begin{aligned}
\|C_0\| &\le \|F'(x_0)^{-1}F'(x^*)\|\|F'(x^*)^{-1}\Big(F'(x_0) - F'(y_0)\Big)\| \\
&\le \frac{\Omega_0(\|x_0 - x^*\|) + \Omega_0(\|y_0 - x^*\|)}{1 - \Omega_0(\|x_0 - x^*\|)} \\
&\le \frac{\Omega_0(\|x_0 - x^*\|) + \Omega_0(\lambda_1(\|x_0 - x^*\|)\|x_0 - x^*\|)}{1 - \Omega_0(\|x_0 - x^*\|)}.
\end{aligned} \tag{2.18}$$

We must show $A_0^{-1} \in L(Y,X)$. By the definition of A_0, the second hypothesis in (H_3), (2.4), (2.8), (2.14), and (2.17), we get that

$$\begin{aligned}\|6(4I-3\alpha C_0)\| &= 6\|F'(x_0)^{-1}\Big((4-3\alpha)F'(x_0)-3\alpha(F'(x_0)-F'(y_0))\Big)\| \\ &\leq 6\left[|4-3\alpha|+\frac{3|\alpha|\Big(\Omega_0(\|x_0-x^*\|)+\Omega_0(\lambda_1(\|x_0-x^*\|)\|x_0-x^*\|)\Big)}{1-\Omega_0(\|x_0-x^*\|)}\right] \\ &= p_1(\|x_0-x^*\|) < p_1(\rho) < 1,\end{aligned}$$

so

$$\|A_0^{-1}\| \leq \frac{1}{1-p_1(\|x_0-x^*\|)}. \tag{2.19}$$

Hence, z_0 is well defined. Moreover, by the second sub-step of method (1.1) for $n=0$, (2.4), (2.6)(for $i=2$), (2.14), and (2.16) - (2.19), we have in turn that

$$\begin{aligned}\|z_0-x^*\| &= \|x_0-x^*-F'(x_0)^{-1}F(x_0)-\frac{3}{4}C_0A_0^{-1}C_0F'(x_0)^{-1}F(x_0)\| \\ &\leq \left[\frac{\int_0^1\Omega((1-\theta)\|x_0-x^*\|)d\theta}{1-\Omega_0(\|x_0-x^*\|)}+\frac{3P(\|x_0-x^*\|)^2\int_0^1\Omega_1(\theta\|x_0-x^*\|)d\theta}{4\big(1-p_1(\|x_0-x^*\|)\big)\big(1-\Omega_0(\|x_0-x^*\|)\big)}\right]\|x_0-x^*\| \\ &= \lambda_2(\|x_0-x^*\|)\|x_0-x^*\| \leq \|x_0-x^*\|,\end{aligned} \tag{2.20}$$

which shows (2.11) for $n=0$ and $z_0 \in \overline{U}(x^*,\rho)$. Next, we must show that $B_0^{-1} \in L(Y,X)$:

$$\begin{aligned}&\|\Big((\beta+\gamma)F'(x^*)\Big)^{-1}\Big[\beta F'(x_0)+\gamma F'(y_0)-(\beta+\gamma)F'(x^*)\Big]\| \\ &\leq |\beta+\gamma|^{-1}\Big[|\beta|\|F'(x^*)^{-1}\Big(F'(x_0)-F'(x^*)\Big)\|+|\gamma|\|F'(x^*)^{-1}\Big(F'(y_0)-F'(x^*)\Big)\|\Big] \\ &\leq |\beta+\gamma|^{-1}\Big[|\beta|\Omega_0(\|x_0-x^*\|)+|\gamma|\Omega_0(\lambda_1(\|x_0-x^*\|)\|x_0-x^*\|)\Big] \\ &= p_2(\|x_0-x^*\|) \leq p_2(\gamma) < 1, (\text{ by}(2.9))\end{aligned}$$

so

$$\|B_0^{-1}F'(x^*)\| \leq \frac{1}{|\beta+\gamma|\Big(1-p_2(\|x_0-x^*\|)\Big)}. \tag{2.21}$$

It also follows that x_1 is well defined. Then, by the last sub-step of method (1.1) for $n=0$, (2.4), (2.6) (for $i=3$), (2.14), (2.16) (for $x_0=z_0$), (2.17), (2.20) and

(2.21), we obtain in turn that

$$\begin{aligned}
\|x_1 - x^*\| &\le \|z_0 - x^*\| + \|B_0^{-1} F'(x^*)\| \Big[\|F'(x^*)^{-1} F'(x_0)\| \\
&\quad + |\delta| \|F'(x^*)^{-1} F(x_0)\| \Big] \|F'(x_0)^{-1} F'(x^*)\| \|F'(x^*)^{-1} F(z_0)\| \\
&\le \left[1 + \frac{\Omega_1(u) + |\delta| \Omega_1(\lambda_1(u)u) \int_0^1 \Omega_1(\theta \lambda_2(u) u) d\theta}{|\beta + \gamma| (1 - p_2(u))(1 - \Omega_0(u))} \right] \|z_0 - x^*\| \\
&\le \lambda_3(u) u \le u, \qquad (2.22)
\end{aligned}$$

where

$$u = \|x_0 - x^*\|,$$

which shows (2.12) for $n = 0$, and $x_1 \in \overline{U}(x^*, \rho)$. The induction is completed by replacing x_0, y_0, z_0, x_1 by x_m, y_m, z_m, x_{m+1} in the preceding estimates. Then, for the inequation

$$\|x_{m+1} - x^*\| \le h \|x_m - x^*\| \le \rho, \qquad (2.23)$$

where

$$h = \lambda_3(\|x_0 - x^*\|) \in [0, 1),$$

we conclude that

$$\lim_{m \longrightarrow +\infty} x_m = x^* \text{ and } x_{m+1} \in \overline{U}(x^*, \rho).$$

Set

$$T = \int_0^1 F'\big(y^* + \theta(x^* - y^*)\big) d\theta \text{ for } y^* \in D_1 \text{ with } F(y^*) = 0.$$

Using (H_2) and (H_5), we get in turn that

$$\|F'(x^*)^{-1} \Big(T - F'(x^*) \Big)\| \le \int_0^1 \Omega_0(\theta \|y^* - x^*\|) d\theta \le \int_0^1 \Omega_0(\theta \rho^*) d\theta < 1,$$

so $T^{-1} \in L(Y, X)$. Finally, from the identity

$$0 = F(x^*) - F(y^*) = T(x^* - y^*),$$

we deduce that $x^* = y^*$.

□

References

[1] Amat, S., Busquier, S. and Gutierrez, J.M., Geometric constructions of iterative functions to solve nonlinear equations, *J. Comput. Appl. Math.* 157, 2003, 197-205.

[2] Argyros, I.K., *Convergence and Applications of Newton-type Iterations*, Springer Verlag, Berlin, Germany, 2008.

[3] Argyros, I.K. and Chen, D., Results on the Chebsyhev method in Banach spaces, *Proyecciones* 12, 1993, 119-128.

[4] Argyros, I.K. and Magreñán, A.A., *A Contemporary Study of Iterative Methods*, Academic Press, Elsevier, San Diego, CA, USA, 2017.

[5] Argyros, I.K., Chen, D. and Qian, Q., A convergence analysis for rational methods with a parameter in Banach space, *Pure Math. Appl.* 5, 1994, 59-73.

[6] Argyros, I.K., On an improved unified convergence analysis for a certain class of Euler-Halley type methods, *J. Korean Soc. Math. Educ. Ser. B Pure Appl. Math.* 13, 2006, 207-215.

[7] Chen, D., Argyros, I.K. and Qian, Q., A local convergence theorem for the Supper-Halley method in a Banach space, *Appl. Math. Lett.* 7, 1994, 49-52.

[8] Ezquerro, J.A. and Hernandez, M.A., Halley's method for operators with unbounded second derivative, *Appl. Numer. Math.* 57, 2007, 354-360.

[9] Ezquerro, J.A. and Hernandez, M.A., New Kantorovich-type conditions for Halley's method, *Appl. Number. Anal. Comput. Math.* 2, 2005, 70-77.

[10] Ezquerro, J.A. and Hernandez, M.A., On the R-order of the Halley method, *J. Math. Anal. Appl.* 303, 2005, 591-601.

[11] Gutierrez, J.M. and Hernandez, M.A., A family of Chebyshev-Halley type methods in Banach space, *Bull. Aust. Math. Soc.* 55, 1997, 113-130.

[12] Gutierrez, J.M. and Hernandez, M.A., An acceleration of Newton's method: Super-Halley method, *Appl. Math. Comput.* 117, 2001, 223-239.

[13] Hernandez, M.A. and Salanova, M.A., A family of Chebyshev-Halley type methods. *Int. J. Comput. Math.* 47, 1993, 59-63.

[14] Huang, Z., On a family of Chebyshev-Halley type methods in Banach space under weaker Smale condition. *Number. Math. JCU* 9, 2000, 37-44.

[15] Kantorovich, L.V. and Akilov, G.P., *Functional Analysis in Normed Spaces*, Pergamon Press, New York, 1964.

Chapter 20

Local Convergence for a Steffensen-Type Method of Order Eight

1. Introduction

The local convergence analysis of method

$$\begin{aligned} y_n &= x_n - A_n - 1F(x_n), \\ z_n &= y_n - B_n A_n^{-1} F(y_n), \\ x_{n+1} &= z_n - C_n A_n^{-1} F(z_n), \end{aligned} \tag{1.1}$$

where

$$\begin{aligned} A_n &= [x_n + G(x_n), x_n; F], \\ B_n &= \frac{13}{4} I - Q_n \left(\frac{7}{2} I - \frac{5}{4} Q_n \right), \\ C_n &= \frac{7}{2} I - Q_n \left(4I - \frac{3}{2} Q_n \right), \\ Q_n &= A_n^{-1} [y_n + G(y_n), y_n; F], \end{aligned}$$

$G : D \subseteq X \to X$ a continuous operator, $[.,.;F] : D \times D \to L(X,X)$ is a divided difference of order one, and $x_0 \in D$ was studied, when $X = \mathbb{R}^k$. The convergence order was shown to be eighth under hypotheses up to the ninth derivative (although these derivatives do not appear in (1.1)) limiting the applicability of the method. As a motivational example, consider $D = \left[-\frac{3}{2}, \frac{1}{2}\right]$ and function F be defined on D by

$$F(x) = \begin{cases} x^3 \ln x^2 + x^5 - x^4, & x \neq 0 \\ 0, & x = 0. \end{cases}$$

Then, clearly $F'''(x)$ is unbounded on D. Hence, the earlier results cannot apply.

In this chapter, we show the local convergence analysis using only the first Fréchet-derivative, and in more general setting of Banach space valued operators. Error estimates and uniqueness results are also provide (not given in earlier works). The order of convergences uses computational order of convergence and approximate order of convergence not requiring high convergence order derivatives. This way, we expand the applicability of method and in the more general setting of a Banach space.

2. Local Convergence

Set $T_0 = [0, \infty)$. Let also $V_0 : T_0 \to T_0$ be a continuous and increasing function with $V_0(0) = 0$. Suppose that equation

$$V_0(t) = 1 \tag{2.1}$$

has a minimal positive solution ρ_0. Set $T = [0, \rho_0)$. Let $V : T \to T_0, V_1 : T \to T_0$ and $V_2 : T \to T_0$ be continuous and increasing functions with $V(0) = 0$. Define functions φ_1 and Ψ_1 on T by

$$\varphi_1(t) = \frac{V(t)}{1 - V_0(t)}$$

and

$$\Psi_1(t) = \varphi_1(t) - 1.$$

We get that $\Psi_1(0) = -1$ and $\Psi_1(t) \to \infty$ as $t \to \rho_0^-$. The intermediate value theorem assures that equation $\Psi_1(t) = 0$ has solutions in $(0, \rho_0)$. Denote by ρ_1 the smallest such solution. Moreover, define functions q, b, φ_2 and Ψ_2 on T by

$$q(t) = \frac{V_1(t)}{1 - W_0(t)},$$

$$b(t) = \frac{13}{4} + q(t)\Big(\frac{7}{2} + \frac{5}{4}q(t)\Big),$$

$$\varphi_2(t) = \Big(1 + \frac{b(t)V_2\big(\varphi_1(t)t\big)}{1 - V_0(t)}\Big)\varphi_1(t)$$

and

$$\Psi_2(t) = \varphi_2(t) - 1.$$

We have that $\Psi_2(0) = -1$ and $\Psi_2(t) \to \infty$ as $t \to \rho_0^-$. Denote by ρ_2 the smallest solution of equation $\Psi_2(t) = 0$ in $(0, \rho_0)$. Furthermore, define functions c, φ_3, and Ψ_3 on T by

$$c(t) = \frac{7}{2} + q(t)\Big(4 + \frac{3}{2}q(t)\Big),$$

$$\varphi_3(t) = \Big(1 + \frac{c(t)V_2\big(\varphi_2(t)t\big)}{1 - V_0(t)}\Big)\Big(1 + \frac{b(t)V_2\big(\varphi_1(t)t\big)}{1 - V_0(t)}\Big)\varphi_1(t)$$

and

$$\Psi_3(t) = \varphi_3(t) - 1.$$

We obtain that $\Psi_3(0) = -1$ and $\Psi_3(t) \to \infty$ as $t \to \rho_0^-$. Denote by ρ_3 the smallest solution of equation $\Psi_3(t) = 0$ in $(0, \rho_0)$.

Finally, define a radius of convergence ρ by

$$\rho = \min\{\rho_i\},\ i = 1, 2, 3. \tag{2.2}$$

Then, we have for each $t \in [0, \rho)$

$$0 \leq V_0(t) < 1, \tag{2.3}$$

$$0 \leq q(t) < 1, \tag{2.4}$$

$$0 \leq b(t) < 1, \tag{2.5}$$

$$0 \leq c(t) < 1, \tag{2.6}$$

and

$$0 \leq \varphi_i(t) < 1. \tag{2.7}$$

The sets $U(x^*, \lambda), \overline{U}(x^*, \lambda)$ denotes the open and closed spheres in X, respectively with center $x^* \in X$ and of radius $\lambda > 0$.
The local convergence analysis uses the conditions (C):

(C_1) $F, G : D \subset X \to X$ are continuously Fréchet - differentiable operators, $[.,.;F] : D \times D \to L(X, X)$ is a divided difference of order one, and there exists $x^* \in D$ such that

$$F(x^*) = G(x^*) = 0 \text{ and } F'(x^*)^{-1} \in L(X, X).$$

(C_2) There exists a function $V_0 : T_0 \to T_0$, continuous and increasing with $V_0(0) = 0$ such that for each $x \in D$.

$$\|F'(x^*)^{-1}\Big([x+G(x), x; F] - F'(x^*)\Big)\| \leq V_0(\|x - x^*\|).$$

Set $D_0 = D \cap U(x^*, \rho_0)$, where ρ_0 is given in (2.1).

(C_3) There exist function $V, V_1, V_2 : T \to T_0$ with $V(0) = 0$ such that for each $x, y \in D_0$.

$$\|F'(x^*)^{-1}\Big([x+G(x), x; F] - [x, x^*; F]\Big)\| \leq V(\|x - x^*\|),$$

$$\|F'(x^*)^{-1}[y+G(y), y; F]\| \leq V_1(\|x - x^*\|),$$

and

$$\|F'(x^*)^{-1}[x, x^*; F]\| \leq V_2(\|x - x^*\|),$$

where

$$y = x - [x+G(x), x; F]^{-1}F(x).$$

(C_4) $\overline{U}(x^*,\rho) \subseteq D$.

(C_5) There exists $\rho^* \geq \rho$ such that

$$V_0(\rho^*) < 1.$$

Set $D_1 = D \cap \overline{U}(x^*,\rho^*)$.

Theorem 2.1 Suppose that the conditions (C) hold. Then, iteration $\{x_n\}$ starting for $x_0 \in U(x^*,\rho) - \{x^*\}$, and produced by method (1.1) converges to the unique solution x^* of equation $F(x) = 0$ in D_1, so that

$$\|y_n - x^*\| \leq \varphi_1(\|x_n - x^*\|)\|x_n - x^*\| \leq \|x_n - x^*\| < \rho, \tag{2.8}$$

$$\|z_n - x^*\| \leq \varphi_2(\|x_n - x^*\|)\|x_n - x^*\| \leq \|x_n - x^*\|, \tag{2.9}$$

and

$$\|x_{n+1} - x^*\| \leq \varphi_3(\|x_n - x^*\|)\|x_n - x^*\| \leq \|x_n - x^*\|, \tag{2.10}$$

where $D_1, r, \varphi_1, \varphi_2$, and φ_3 are defined previously.

Proof. Inequations (2.8) - (2.10) shall be shown using mathematical induction. It follows from the condition $x_0 \in U(x^*,\rho) - \{x^*\}$, (2.1), (2.2), (C_1) and (C_2) that

$$\|F'(x^*)\Big([x_0 + G(x_0), x_0; F] - F'(x^*)\Big)\| \leq V_0(\|x_0 - x^*\|) \leq V(\rho) < 1. \tag{2.11}$$

The Banach lemma on invertible operators [15], and (2.11) imply the existence of A_0^{-1}, and

$$\|A_0^{-1}F'(x^*)\| \leq \frac{1}{1 - V_0(\|x_0 - x^*\|)}. \tag{2.12}$$

We also have that y_0, z_0, x_1 are well defined by method (1.1) for $n = 0$, since A_0^{-1} exists. We can write

$$\begin{aligned} y_0 - x^* &= x_0 - x^* - A_0^{-1}F(x_0) \\ &= A_0^{-1}\Big(A_0 - [x_0, x^*, F]\Big)(x_0 - x^*). \end{aligned} \tag{2.13}$$

Using (2.2), (2.3), (2.7)(for $i = 1$), (2.12), (2.13) (C_1), and the first condition in (C_3), we obtain in turn that

$$\begin{aligned}\|y_0 - x^*\| &\leq \|A_0^{-1}F'(x^*)\|\|F'(x^*)^{-1}\Big(A_0 - [x_0,x^*;F]\Big)\|\|x_0 - x^*\| \\ &\leq \frac{V(\|x_0 - x^*\|)}{1 - V_0(\|x_0 - x^*\|)}\|x_0 - x^*\| \\ &\leq \varphi_1(\|x_0 - x^*\|)\|x_0 - x^*\| \leq \|x_0 - x^*\| < \rho, \end{aligned} \tag{2.14}$$

which shows (2.8) for $n = 0$, and $y_0 \in U(x^*, \rho)$.
By (C_1), the second condition in (C_2), (2.2), and (2.4), we get in turn that

$$F(y_0) = F(y_0) - F(x^*) = [y_0, x^*; F](x_0 - x^*), \tag{2.15}$$

so

$$\begin{aligned}\|F'(x^*)^{-1}F(y_0)\| &= \|F'(x^*)^{-1}[y_0,x^*;F](y_0 - x^*)\| \\ &\leq V_2(\|y_0 - x^*\|)d\theta\|y_0 - x^*\| \\ &\leq V_2(\varphi_1(\|x_0 - x^*\|)\|x_0 - x^*\|)\varphi_1(\|x_0 - x^*\|)\|x_0 - x^*\|. \end{aligned} \tag{2.16}$$

We need an upper bound on $\|Q_0\|$ and $\|B_0\|$:

$$\begin{aligned}\|Q_0\| &\leq \|A_0^{-1}F'(x^*)\|\|F'(x^*)^{-1}[y_0 + G(y_0), y_0; F]\| \\ &\leq \frac{V_1(\|x_0 - x^*\|)}{1 - V_0(\|x_0 - x^*\|)} = q(\|x_0 - x^*\|) \end{aligned} \tag{2.17}$$

and

$$\begin{aligned}\|B_0\| &\leq \frac{13}{4} + \|Q_0\|\Big(\frac{7}{2} + \frac{5}{4}\|Q_0\|\Big) \\ &\leq \frac{13}{4} + q(\|x_0 - x^*\|)\Big(\frac{7}{2} + \frac{5}{4}q(\|x_0 - x^*\|)\Big) = b(\|x_0 - x^*\|). \end{aligned} \tag{2.18}$$

Then, by (2.2), (2.5), (2.7)(for $i = 2$), (2.12), (2.14), (2.16) and (2.18), we have in turn that

$$\begin{aligned}
\|z_0 - x^*\| &= \|y_0 - x^* - B_0 A_0^{-1} F(y_0)\| \\
&\leq \|y_0 - x^*\| + \frac{b(\|x_0 - x^*\|) V_2(\|y_0 - x^*\|)\|y_0 - x^*\|}{1 - V_0(\|x_0 - x^*\|)} \\
&\leq \left(1 + \frac{b(\|x_0 - x^*\|) V_2\big(\varphi_1(\|x_0 - x^*\|)\|x_0 - x^*\|\big)}{1 - V_0(\|x_0 - x^*\|)}\right) \varphi_1(\|x_0 - x^*\|)\|x_0 - x^*\| \\
&= \varphi_2(\|x_0 - x^*\|)\|x_0 - x^*\| \leq \|x_0 - x^*\|, \qquad (2.19)
\end{aligned}$$

which shows (2.9) for $n = 0$ and $z_0 \in U(x^*, \rho)$. We need an upper bound on $\|C_0\|$:

$$\begin{aligned}
\|C_0\| &\leq \frac{7}{2} + \|Q_0\| \left(4 + \frac{3}{2}\|Q_0\|\right) \\
&\leq \frac{7}{2} + q(\|x_0 - x^*\|)\left(4 + \frac{3}{2} q(\|x_0 - x^*\|)\right) = c(\|x_0 - x^*\|). \qquad (2.20)
\end{aligned}$$

Then, by (2.2), (2.6), (2.7)(for $(i = 3)$), (2.16)(for $y_0 = z_0$), (2.12), (2.19) and (2.20), we obtain in turn that

$$\begin{aligned}
\|x_1 - x^*\| &\leq \|z_0 - x^*\| + \frac{c(u) V_2\big(\varphi_2(u)u\big)\|z_0 - x^*\|}{1 - V_0(u)} \\
&\leq \left(1 + \frac{c(u) V_2\big(\varphi_2(u)u\big)}{1 - V_0(u)}\right)\left(1 + \frac{b(u) V_2\big(\varphi_1(u)u\big)}{1 - V_0(u)}\right)\varphi_1(u)u \\
&= \varphi_3(u)u \leq u, \qquad (2.21)
\end{aligned}$$

where

$$u = \|x_0 - x^*\|,$$

which shows (2.10) for $n = 0$ and $x_1 \in U(x^*, \rho)$. Then, replace x_0, y_0, z_0, x_1 by x_m, y_m, z_m, x_{m+1} in the preceding inequations to finish the induction for (2.8) - (2.10). In view of the inequation

$$\|x_{m+1} - x^*\| \leq p\|x_m - x^*\| < \rho, \qquad (2.22)$$

which shows

$$\lim_{m \to +\infty} x_m = x^* \text{ and } x_{m+1} \in U(x^*, \rho).$$

Let $H = [y^* + G(y^*), x^*; F]$ for some $y^* \in D_1$ with $F(y^*) = G(y^*) = 0$. Using (C_2) and (C_5), we get that

$$\begin{aligned}\|F'(x^*)^{-1}\Big(H - F'(x^*)\Big)\| &\leq V_0(\|y^* - x^*\|)\\ &\leq V_0(\rho^*) < 1,\end{aligned}$$

so $H^{-1} \in L(X, X)$. Then we can write by identity

$$0 = F(y^*) - F(x^*) = L(y^* - x^*),$$

leading to $x^* = y^*$.

□

Remark 2.2 We shall show how to choose ρ_0, V_0, V, and V_2 in some interesting special cases. Suppose that the following conditions hold:

$$\begin{aligned}&\|F'(x^*)^{-1}\Big([x + G(x), x; F] - F'(x^*)\Big)\|\\ &\qquad \leq \lambda_1\|x + G(x) - G(x^*) - x^*\| + \lambda_2\|x - x^*\|\\ &\|F'(x^*)^{-1}\Big([x + G(x), x; F] - [x, x^*; F]\Big)\|\\ &\qquad \leq \lambda_3\|x + G(x) - x^*\| + \lambda_4\|x - x^*\|\\ &\|F'(x^*)^{-1}[y + G(y), y; F]\| \leq \lambda_5\\ &\|F'(x^*)^{-1}[x, x^*; F]\| \leq \lambda_6\end{aligned}$$

and

$$\|G(x) - G(x^*)\| \leq \mu\|x - x^*\|.$$

Then, we can choose:

$$\begin{aligned}&V_0(t) = [\lambda_1(1 + \mu) + \lambda_2]t, \rho_0 = \frac{1}{(1 + \mu)\lambda_1 + \lambda_2},\\ &V(t) = (\lambda_3\mu + \lambda_4)t,\\ &V_1(t) = \lambda_5, \text{ and } V_2(t) = \lambda_6.\end{aligned}$$

References

[1] Amat, S., Busquier, S. and Gutierrez, J.M., Geometric constructions of iterative functions to solve nonlinear equations, *J. Comput. Appl. Math.* 157, 2003, 197-205.

[2] Argyros, I.K., *Convergence and Applications of Newton-type Iterations*, Springer Verlag, Berlin, Germany, 2008.

[3] Argyros, I.K. and Chen, D., Results on the Chebsyhev method in Banach spaces, *Proyecciones* 12, 1993, 119-128.

[4] Argyros, I.K. and Magreñán, A.A., *A Contemporary Study of Iterative Methods*, Academic Press, Elsevier, San Diego, CA, USA, 2017.

[5] Argyros, I.K., Chen, D. and Qian, Q., A convergence analysis for rational methods with a parameter in Banach space, *Pure Math. Appl.* 5, 1994, 59-73.

[6] Argyros, I.K., On an improved unified convergence analysis for a certain class of Euler-Halley type methods, *J. Korean Soc. Math. Educ. Ser. B Pure Appl. Math.* 13, 2006, 207-215.

[7] Chen, D., Argyros, I.K. and Qian, Q., A local convergence theorem for the Supper-Halley method in a Banach space, *Appl. Math. Lett.* 7, 1994, 49-52.

[8] Ezquerro, J.A. and Hernandez, M.A., Halley's method for operators with unbounded second derivative, *Appl. Numer. Math.* 57, 2007, 354-360.

[9] Ezquerro, J.A. and Hernandez, M.A., New Kantorovich-type conditions for Halley's method, *Appl. Number. Anal. Comput. Math.* 2, 2005, 70-77.

[10] Ezquerro, J.A. and Hernandez, M.A., On the R-order of the Halley method, *J. Math. Anal. Appl.* 303, 2005, 591-601.

[11] Gutierrez, J.M. and Hernandez, M.A., A family of Chebyshev-Halley type methods in Banach space, *Bull. Aust. Math. Soc.* 55, 1997, 113-130.

[12] Gutierrez, J.M. and Hernandez, M.A., An acceleration of Newton's method: Super-Halley method, *Appl. Math. Comput.* 117, 2001, 223-239.

[13] Hernandez, M.A. and Salanova, M.A., A family of Chebyshev-Halley type methods. *Int. J. Comput. Math.* 47, 1993, 59-63.

[14] Huang, Z., On a family of Chebyshev-Halley type methods in Banach space under weaker Smale condition. *Number. Math. JCU* 9, 2000, 37-44.

[15] Kantorovich, L.V. and Akilov, G.P., *Functional Analysis in Normed Spaces*, Pergamon Press, New York, 1964.

Chapter 21

Convergence of a Sixth Order Method

1. Introduction

The local convergence analysis of method

$$
\begin{aligned}
y_n &= x_n - \frac{1}{2}F'(x_n)^{-1}F(x_n),\\
z_n &= \frac{1}{3}(4y_n - x_n),\\
u_n &= y_n + A_n^{-1}F(x_n),\\
v_m &= u_n + 2A_n^{-1}F(u_n),\\
x_{n+1} &= v_n + 2A_n^{-1}F(v_n),
\end{aligned}
\tag{1.1}
$$

where

$$A_n = F'(x_n) - 3F'(z_n)$$

and $x_0 \in D$ was studied, when $X = Y = \mathbb{R}^k$. The convergence order was shown to be six under hypotheses up to the seventh derivative (although these derivatives do not appear in (1.1)) limiting the applicability of the method. As a motivational example, consider $D = \left[-\frac{3}{2}, \frac{1}{2}\right]$ and function F be defined on D by

$$f(x) = \begin{cases} x^3 \ln x^2 + x^5 - x^4, & x \neq 0 \\ 0, & x = 0. \end{cases}$$

Then, clearly $F'''(x)$ is unbounded on D. Hence, the earlier results cannot apply.

In this chapter, we show the local convergence analysis using only the first Fréchet-derivative, and in more general setting of Banach space valued operators. Error estimates and uniqueness results are also provide (not given in earlier works). The order of convergences uses computational order of convergence and approximate order of convergence not requiring high convergence order derivatives. This way, we expand the applicability of method and in the more general setting of a Banach space.

2. Local Convergence

Let $\varphi_0 : \mathbb{R}_+ \cup \{0\} \to \mathbb{R}_+ \cup \{0\}$ be a continuous and increasing function with $\varphi_0(0) = 0$.
Suppose that equation

$$\varphi_0(t) = 1 \tag{2.1}$$

has a minimal positive solution r_0. Let $I_0 = [0, r_0)$. Let functions $\varphi : I_0 \to \mathbb{R}_+ \cup \{0\}, \varphi_1 : I_0 \to \mathbb{R}_+ \cup \{0\}$ be continuous and increasing functions with $\varphi(0) = 0$.
Define functions Ψ_1 and $\overline{\Psi}_1$ on I_0 by

$$\Psi_1(t) = \frac{\int_0^1 \varphi\Big((1-\theta)t\Big)d\theta + \frac{1}{2}\int_0^1 \varphi_1\Big((\theta t)d\theta\Big)}{1-\varphi_0(t)}$$

and

$$\overline{\Psi}(t) = \Psi_1(t) - 1.$$

Suppose that

$$\frac{\varphi_1(0)}{2} - 1 < 0. \tag{2.2}$$

We have by (2.2), $\overline{\Psi}_1(0) < 0$ and $\overline{\Psi}_1(t) \to \infty$ as $t \to r_0^-$. Then, equation $\overline{\Psi}_1(t) = 0$ has at least one solution in $(0, r_0)$ by the intermediate value theorem. Denote by r_1 the smallest such solution. Define also functions Ψ_2 and $\overline{\Psi}_2$ on I_0 by

$$\Psi_2(t) = \Psi_1(t) + \frac{1}{6}\frac{\int_0^1 \varphi_1(\theta t)d\theta}{1-\varphi_0(t)}$$

and

$$\overline{\Psi}_2(t) = \Psi_2(t) - 1.$$

Suppose that

$$\frac{2}{3}\varphi_1(0) - 1 < 0. \tag{2.3}$$

We get by (2.3) $\overline{\Psi}_2 < 0$ and $\overline{\Psi}_2(t) \to \infty$ as $t \to r_0^-$. Denote by r_2 the smallest solution of equation $\overline{\Psi}_2(t) = 0$ in $(0, r_0)$.
Define functions p and q on I_0 by

$$p(t) = \frac{1}{2}[\varphi_0(t) + 3\varphi_0(\Psi_2(t)t)].$$

Suppose equation $p(t) = 1$ has a minimal positive solution ρ. Moreover, suppose that equation

$$\varphi_0(\Psi_3(t)t) = 1 \tag{2.4}$$

has a minimal solutions in $(r_0, \rho_1), \rho_1 = \min\{r_1, \rho\}$ denoted by ρ_2. Furthermore, define functions Ψ_4 and $\overline{\Psi}_4$ on $I_1 := [0, \rho_2)$ by

$$\Psi_4(t) = \left[\frac{\int_0^1 \varphi((1-\theta)\Psi_3(t)t)d\theta}{1 - \varphi_0(\Psi_3(t)t)} + \frac{\left(\varphi_0(t) + 3\varphi_0(\Psi_2(t)t) + 2\varphi_0(\Psi_3(t)t)\right)}{2(1 - \varphi_0(\Psi_3(t)t))(1 - p(t))}\right]\Psi_3(t)$$

and

$$\overline{\Psi}_4(t) = \Psi_4(t) - 1.$$

We get $\overline{\Psi}_4(0) = -1$ and $\overline{\Psi}_4(t) \to \infty$ as $t \to \rho_3^-$, where $\rho_3 = \min\{r_0, \rho, \rho_1\}$. Denote by r_4 the smallest solution of equation $\overline{\Psi}_4(t) = 0$.
Suppose that equation

$$\varphi_0(\Psi_4(t)t) = 1 \tag{2.5}$$

has a minimal solutions in $(0, \rho_3)$ denote by ρ_4. Furthermore, define functions Ψ_5 and $\overline{\Psi}_5$ on $I_2 := [0, \rho_4)$ by

$$\Psi_5(t) = \left[\frac{\int_0^1 \varphi((1-\theta)\Psi_4(t)t)d\theta}{1 - \varphi_0(\Psi_4(t)t)} + \frac{\left(\varphi_0(t) + 3\varphi_0(\Psi_2(t)t) + 2\varphi_0(\Psi_4(t)t)\right)\int_0^1 \varphi_1(\theta\Psi_4(t)t)d\theta}{2(1 - \varphi_0(\Psi_4(t)t))(1 - p(t))}\right]\Psi_4(t)$$

and

$$\overline{\Psi}_5(t) = \Psi_5(t) - 1,$$

where $\rho_4 = \min\{r_0, \rho, \rho_2, \rho_3\}$.
We get $\overline{\Psi}_5(0) = -1 < 0$ and $\overline{\Psi}_5(t) \to \infty$ as $t \to \rho_4^-$. Denote by r_5 the smallest such solution of equation $\overline{\Psi}_5(t) = 0$ in $(0, \rho_4)$. Finally, define a radius of convergence r by

$$r = \min\{r_i\},\ i = 1,2,3,4,5. \tag{2.6}$$

It follows that for each $t \in [0, r)$

$$0 \leq \varphi_0(t) < 1, \tag{2.7}$$

$$0 \leq p(t) < 1, \tag{2.8}$$

and

$$0 \leq \Psi_i(t) < 1. \tag{2.9}$$

Sets $U(x,a)$ and $\overline{U}(x,a)$ stand, respectively for open and closed balls in X with center $x \in X$ and of radius $a > 0$.
The local convergence analysis of method (1.1) utilizes the conditions (A):

(A_1) $F : D \subset X \to Y$ is a continuously Fréchet-differentiable operator and there exist $x^* \in D$ such that

$$F(x^*) = 0 \text{ and } F'(x^*)^{-1} \in L(Y,X).$$

(A_2) There exists functions $\varphi_0 : \mathbb{R}_+ \cup \{0\} \to \mathbb{R}_+ \cup \{0\}$ continuous, increasing with $\varphi_0(0) = 0$ such that for each $x \in D$

$$\|F'(x^*)^{-1}\Big(F'(x) - F'(x^*)\Big)\| \leq \varphi_0(\|x - x^*\|).$$

Set $D_0 = D \cap U(x^*, r_0)$, where r_0 is given in (2.1).

(A_3) There exists functions $\varphi : I_0 \to \mathbb{R}_+ \cup \{0\}, \varphi_1 : I_0 \to \mathbb{R}_+ \cup \{0\}$ continuous, increasing with $\varphi(0) = 0$ such that for each $x, y \in D_0$

$$\|F'(x^*)^{-1}\Big(F'(x) - F'(y)\Big)\| \leq \varphi(\|x - y\|)$$

and

$$\|F'(x^*)^{-1} F'(x)\| \leq \varphi_1(\|x - x^*\|).$$

(A_4) $\overline{U}(x^*,r) \subset D$, (2.3) - (2.5) hold.

(A_5) There exists $r^* \geq r$ such that

$$\int_0^1 \varphi_0(\theta r^*)d\theta < 1.$$

Set $D_1 = D \cap U(x^*,r^*)$.

Theorem 2.1 Under the condition (A), iteration $\{x_n\}$ for $x_0 \in U(x^*,r) - \{x^*\}$ produced by method (1.2(1.1)) converges to the only solution x^* of equation $F(x) = 0$ in D_1, and

$$\|y_n - x^*\| \leq \Psi_1(\|x_n - x^*\|)\|x_n - x^*\| \leq \|x_n - x^*\| < r, \tag{2.10}$$

$$\|z_n - x^*\| \leq \Psi_2(\|x_n - x^*\|)\|x_n - x^*\| \leq \|x_n - x^*\|, \tag{2.11}$$

$$\|u_n - x^*\| \leq \Psi_3(\|x_n - x^*\|)\|x_n - x^*\| \leq \|x_n - x^*\|, \tag{2.12}$$

$$\|v_n - x^*\| \leq \Psi_4(\|x_n - x^*\|)\|x_n - x^*\| \leq \|x_n - x^*\|, \tag{2.13}$$

and

$$\|x_{n+1} - x^*\| \leq \Psi_5(\|x_n - x^*\|)\|x_n - x^*\| \leq \|x_n - x^*\|, \tag{2.14}$$

where $D_1, \Psi_i, i = 1,2,3,4,5$ and r are defined previously.

Proof. Error bounds (2.10) - (2.14) are shown using mathematical induction. By condition $x_0 \in U(x^*,r) - \{x^*\}$, (2.7), and (A_2), we obtain in turn that

$$\|F'(x^*)^{-1}\Big(F'(x_0) - F'(x^*)\Big)\| \leq \varphi_0(\|x_0 - x^*\|) \leq \varphi_0(r) < 1, \tag{2.15}$$

so $F'(x_0)^{-1} \in L(Y,X)$ by the Banach lemma on invertible operators [15], and

$$\|F'(x_0)^{-1}F'(x^*)\| \leq \frac{1}{1 - \varphi_0(\|x_0 - x^*\|)}. \tag{2.16}$$

It also follows by the first and second sub-step of method (1.1) that y_0 and z_0 are well defined. Using the first sub-step of method (1.1) for $n = 0$, we can write

$$y_0 - x^* = x_0 - x^* - F'(x_0)^{-1}F(x_0) + \frac{1}{2}F'(x_0)^{-1}F(x_0). \tag{2.17}$$

We have by (A_1) and (A_3) that

$$F(x_0) = F(x_0) - F(x^*) = \int_0^1 F'(\theta\|x_0 - x^*\|)d\theta\|x_0 - x^*\|, \tag{2.18}$$

so

$$\|F'(x^*)^{-1}F(x_0)\| \leq \int_0^1 \varphi_1\Big(\theta\|x_0 - x^*\|\Big)d\theta\|x_0 - x^*\|, \tag{2.19}$$

where, we used that $x^* + \theta(x_0 - x^*) \in U(x^*, r)$, since $\|x^* + \theta(x_0 - x^*)\| = \theta\|x_0 - x^*\| < r$. Then, by (2.17), (2.9), (2.19)(for $i = 1$), (2.6), we have in turn that

$$\begin{aligned}
\|y_0 - x^*\| &\leq \|F'(x_0)^{-1}F'(x^*)\|\|\int_0^1 F'(x^*)^{-1}\Big(F'(x^* + \theta(x_0 - x^*))\\
&- F'(x_0)\Big)(x_0 - x^*)d\theta\| + \frac{1}{2}\|F'(x_0)^{-1}F'(x^*)\|\|F'(x^*)^{-1}F(x_0)\|\\
&\leq \frac{\int_0^1 \varphi((1-\theta)\|x_0 - x^*\|)d\theta\|x_0 - x^*\| + \frac{1}{2}\int_0^1 \varphi_1(\theta\|x_0 - x^*\|)d\theta\|x_0 - x^*\|}{1 - \varphi_0(\|x_0 - x^*\|)}\\
&= \Psi_1(\|x_0 - x^*\|)\|x_0 - x^*\| \leq \|x_0 - x^*\| < r,
\end{aligned} \tag{2.20}$$

which shows (2.10) for $n = 0$ and $y_0 \in U(x^*, r)$. By the second sub-step of method (1.1), (2.6), (2.9)(for $i = 2$), and (2.20), we have in turn that

$$\begin{aligned}
\|z_0 - x^*\| &= \|y_0 - x^* + \frac{1}{3}(y_0 - x^*)\|\\
&\leq \|y_0 - x^*\| + \frac{1}{3}\|y_0 - x^*\|\\
&\leq \Psi_1(\|x_0 - x^*\|)\|x_0 - x^*\| + \frac{1}{3}\frac{\int_0^1 \varphi_1(\theta\|x_0 - x^*\|)d\theta\|x_0 - x^*\|}{1 - \varphi_0(\|x_0 - x^*\|)}\\
&= \Psi_1(\|x_0 - x^*\|)\|x_0 - x^*\| \leq \|x_0 - x^*\|,
\end{aligned} \tag{2.21}$$

which shows (2.11) for $n = 0$ and $z_0 \in U(x^*, r)$. We must show that $A_0^{-1} \in$

$L(Y,X)$. Using (2.6), and (A_2), we get in turn that

$$\begin{aligned}
&\|(-2F'(x^*))^{-1}\Big(A_0+2F'(x^*)\Big)\| \\
&\leq \frac{1}{2}\|F'(x^*)^{-1}\Big[\big(F'(x_0)-F'(x^*)\big)+3\big(F'(x^*)-F'(z_0)\big)\Big]\| \\
&\leq \frac{1}{2}\Big[\|F'(x^*)^{-1}\Big(F'(x_0)-F'(x^*)\Big)\|+3\|F'(x^*)^{-1}\Big(F'(z_0)-F'(x^*)\Big)\|\Big] \\
&\leq \frac{1}{2}\Big[\varphi_0(\|x_0-x^*\|)+3\varphi_0(\|z_0-x^*\|)\Big] \\
&\leq \frac{1}{2}\Big(\varphi_0(\|x_0-x^*\|)+3\varphi_0(\Psi_2(\|x_0-x^*\|)\|x_0-x^*\|)\Big) \\
&= p(\|x_0-x^*\|)\leq P(r)<1,
\end{aligned} \tag{2.22}$$

so $A_0^{-1}\in L(Y,X)$, and

$$\|A_0^{-1}F'(x^*)\|\leq \frac{1}{2\big(1-P(\|x_0-x^*\|)\big)}. \tag{2.23}$$

It also follows that u_0, v_0 and x_1 are well defined by the third, fourth. and fifth sub-step of method (1.1) for $n=0$, respectively.
Moreover, we have by (2.6), (2.9)(for $i=3,4,5$), (2.16), (2.19)(for $x_0=u_0,v_0$), (2.20), (2.21), and (2.23) that

$$\begin{aligned}
\|y_0-x^*\| &= \|y_0-x^*+F'(x_0)^{-1}\Big[\frac{1}{2}\Big(F'(x_0)-3F'(z_0)\Big) \\
&\qquad +F'(x_0)\Big]A_0^{-1}F(x_0)\| \\
&= \|y_0-x^*+\frac{3}{2}F'(x_0)^{-1}\Big(F'(x_0)-F'(z_0)\Big)A_0^{-1}F(x_0)\| \\
&\leq \Psi_1(\|x_0-x^*\|)\|x_0-x^*\| \\
&+\frac{3\Big(\varphi_0(\|x_0-x^*\|)+\varphi_0(\|z_0-x^*\|)\Big)\int_0^1\varphi_1(\theta\|x_0-x^*\|)d\theta\|x_0-x^*\|}{4\big(1-\varphi_0(\|x_0-x^*\|)\big)\big(1-p(\|x_0-x^*\|)\big)} \\
&\leq \Psi_3(\|x_0-x^*\|)\|x_0-x^*\|\leq\|x_0-x^*\|,
\end{aligned} \tag{2.24}$$

which shows (2.12) for $n = 0$ and $u_0 \in U(x^*, r)$, we also have from the fourth and fifth sub-steps of method (1.1) for $n = 0$ that

$$\begin{aligned} v_0 - x^* &= u_0 - x^* - F(u_0)^{-1}F(u_0) \\ &\quad + F'(u_0)^{-1}\Big[\Big(F'(x_0) - F'(x^*)\Big) - 3\Big(F'(z_0) - F'(x^*)\Big) \\ &\qquad + 2\Big(F'(u_0) - F'(x^*)\Big)\Big]A_0^{-1}F(u_0) \end{aligned}$$

so

$$\begin{aligned} \|v_0 - x^*\| &\leq \Bigg[\frac{\int_0^1 \varphi\big((1-\theta)s\big)d\theta}{1 - \varphi_0(s)} \\ &\quad + \frac{\Big(\varphi_0(\|x_0 - x^*\|) + 3\varphi_0(\|z_0 - x^*\|) + 2\varphi_0(s)\Big)\int_0^1 \varphi_1(\theta s)d\theta}{2\big(1 - \varphi_0(s)\big)\big(1 - p(\|x_0 - x^*\|)\big)}\Bigg]s \\ &\leq \Psi_4(\|x_0 - x^*\|)\|x_0 - x^*\| \leq \|x_0 - x^*\|, \end{aligned} \tag{2.25}$$

where

$$s = \|u_0 - x^*\|,$$

and

$$\begin{aligned} x_1 - x^* = v_0 - x^* - F'(v_0)^{-1}F(v_0) + F'(v_0)^{-1}\Big[\Big(F'(x_0) - F'(x^*)\Big) - 3\Big(F'(z_0) - F'(x^*)\Big) \\ + 2\Big(F'(v_0) - F'(x^*)\Big)\Big]A_0^{-1}F(v_0) \end{aligned}$$

so

$$\begin{aligned} \|x_1 - x^*\| &\leq \Bigg[\frac{\int_0^1 \varphi\big((1-\theta)b\big)d\theta}{1 - \varphi_0(b)} \\ &\quad + \frac{\Big(\varphi_0(\|x_0 - x^*\|) + 3\varphi_0(\|z_0 - x^*\|) + 2\varphi_0(b)\Big)\int_0^1 \varphi_1(\theta b)d\theta}{2\big(1 - \varphi_0(b)\big)\big(1 - p(\|x_0 - x^*\|)\big)}\Bigg]b \\ &\leq \Psi_5(\|x_0 - x^*\|)\|x_0 - x^*\| \leq \|x_0 - x^*\|, \end{aligned} \tag{2.26}$$

where

$$b = \|v_0 - x^*\|$$

which show (2.13), (2.14), respectively, $v_0 \in U(x^*, r)$ and $x_1 \in U(x^*, r)$, where we also used (2.16) (for $x_0 = u_0, v_0$), (2.6), (2.9) (for $i = 4, 5$), (2.18) (for $x_0 = u_0, v_0$), (2.20), (2.21), and (2.23) - (2.25).

The induction is finished by replacing $x_0, y_0, z_0, u_0, v_0, x_1$ by $x_m, y_m, z_m, u_m, v_m, x_{m+1}$ in the previous results. It then follows from the result

$$\|x_{m+1} - x^*\| \leq h\|x_m - x^*\| < r, \tag{2.27}$$

where $h = \Psi_5(\|x_0 - x^*\|) \in [0, 1)$ that

$$\lim_{m \to +\infty} x_m = x^* \text{ and } x_{m+1} \in U(x^*, r).$$

Let

$$Q = \int_0^1 F'\Big(x^* + \theta(y^* - x^*)\Big)d\theta \text{ for } y^* \in D_1 \text{ with } F(y^*) = 0.$$

Then, by (A_5) and (A_2), we get that

$$\|F'(x^*)^{-1}\Big(Q - F'(x^*)\Big)\| \leq \int_0^1 \varphi_0(\theta\|x^* - y^*\|)d\theta \leq \int_0^1 \varphi_0(\theta r^*)d\theta < 1, \tag{2.28}$$

so $Q^{-1} \in L(Y, X)$. Finally, from identity

$$0 = F(y^*) - F(x^*) = Q(y^* - x^*),$$

we deduce that $x^* = y^*$

□

References

[1] Amat, S., Busquier, S. and Gutierrez, J.M., Geometric constructions of iterative functions to solve nonlinear equations, *J. Comput. Appl. Math.* 157, 2003, 197-205.

[2] Argyros, I.K., *Convergence and Applications of Newton-type Iterations*, Springer Verlag, Berlin, Germany, 2008.

[3] Argyros, I.K. and Chen, D., Results on the Chebsyhev method in Banach spaces, *Proyecciones* 12, 1993, 119-128.

[4] Argyros, I.K. and Magreñán, A.A., *A Contemporary Study of Iterative Methods*, Academic Press, Elsevier, San Diego, CA, USA, 2017.

[5] Argyros, I.K., Chen, D. and Qian, Q., A convergence analysis for rational methods with a parameter in Banach space, *Pure Math. Appl.* 5, 1994, 59-73.

[6] Argyros, I.K., On an improved unified convergence analysis for a certain class of Euler-Halley type methods, *J. Korean Soc. Math. Educ. Ser. B Pure Appl. Math.* 13, 2006, 207-215.

[7] Chen, D., Argyros, I.K. and Qian, Q., A local convergence theorem for the Supper-Halley method in a Banach space, *Appl. Math. Lett.* 7, 1994, 49-52.

[8] Ezquerro, J.A. and Hernandez, M.A., Halley's method for operators with unbounded second derivative, *Appl. Numer. Math.* 57, 2007, 354-360.

[9] Ezquerro, J.A. and Hernandez, M.A., New Kantorovich-type conditions for Halley's method, *Appl. Number. Anal. Comput. Math.* 2, 2005, 70-77.

[10] Ezquerro, J.A. and Hernandez, M.A., On the R-order of the Halley method, *J. Math. Anal. Appl.* 303, 2005, 591-601.

[11] Gutierrez, J.M. and Hernandez, M.A., A family of Chebyshev-Halley type methods in Banach space, *Bull. Aust. Math. Soc.* 55, 1997, 113-130.

[12] Gutierrez, J.M. and Hernandez, M.A., An acceleration of Newton's method: Super-Halley method, *Appl. Math. Comput.* 117, 2001, 223-239.

[13] Hernandez, M.A. and Salanova, M.A., A family of Chebyshev-Halley type methods. *Int. J. Comput. Math.* 47, 1993, 59-63.

[14] Huang, Z., On a family of Chebyshev-Halley type methods in Banach space under weaker Smale condition. *Number. Math. JCU* 9, 2000, 37-44.

[15] Kantorovich, L.V. and Akilov, G.P., *Functional Analysis in Normed Spaces*, Pergamon Press, New York, 1964.

Chapter 22

Roots of Polynomials and Their Applications

1. Introduction

We introduce a special class of recurrent polynomials f_n $(n \geq 1)$ of degree n with real coefficients.Then, we find sufficient conditions under which each polynomial f_n has a unique positive root s_n, such that $s_{n+1} \leq s_n$ $(n \geq 1)$. The first root s_1, as well as the last one denoted by s_∞ are expressed in simple closed form.

Two applications are provided. In the first one, we show how to use s_1 and s_∞ to locate any s_n belonging in $(s_\infty, s_1]$ $(n \geq 1)$.In the second one, using this technique on Newton's method (2.19), we show that the famous for its simplicity and clarity Newton-Kantorovich condition (3.2) for solving equations can always replaced by a weaker one (2.16).

Moreover, the ratio of the quadratic convergence of Newton's method $2q_0$ (see, (2.16), (2.24), and (3.2)) under our approach is smaller than $2q_K$ given in [2].

2. Locating Roots of Polynomials

We need the main result on locating roots of polynomials.

Theorem 2.1 Let $a > 0, b > 0$, and $c < 0$ be given constants. Define polynomials f_n $(n \geq 1)$, g on $[0, \infty)$ by

$$f_n(s) = bs^n + as^{n-1} + b(s^{n-1} + s^{n-2} + \cdots + 1) + c, \tag{2.1}$$

and

$$g(s) = bs^2 + as - a. \tag{2.2}$$

Set

$$d = \frac{2a}{a + \sqrt{a^2 + 4ab}}. \tag{2.3}$$

Assume

$$d \leq 1 + \frac{b}{c}, \tag{2.4}$$

and

$$a + b + c < 0. \tag{2.5}$$

Then, each polynomial $f_n(n \geq 1)$ has a unique positive root s_n.
Moreover, the following estimates hold for all $n \geq 1$

$$1 + \frac{b}{c} \leq s^* \leq s_{n+1} \leq s_n, \tag{2.6}$$

and

$$f_n(d) \leq 0, \tag{2.7}$$

where,

$$s^* = \lim_{n \to \infty} s_n.$$

Proof. Each polynomial f_n has a unique positive root $s_n(n \geq 1)$, by the Descarte's rule of signs. Polynomial f_n can be written for $s \in [0, 1)$

$$f_n(s) = bs^n + as^{n-1} + b\frac{1 - s^n}{1 - s} + c. \tag{2.8}$$

By letting $n \to \infty$, we get

$$f_\infty(s) = \lim_{n\to\infty} f_n(s) = \frac{b}{1-s} + c. \tag{2.9}$$

Function f_∞ has a unique positive root denoted by s_∞, and given by

$$s_\infty = 1 + \frac{b}{c} < 1. \tag{2.10}$$

Function f_∞ is also increasing, since

$$f'_\infty(s) = \frac{b}{(1-s)^2} > 0. \tag{2.11}$$

Furthermore, we shall show estimates (2.6), and (2.7) hold.

We need the relationship between two consecutive polynomials $f_n's(n \geq 1)$

$$\begin{aligned} f_{n+1}(s) &= bs^{n+1} + as^n + b(s^n + \cdots + 1) + c \\ &= bs^n + as^{n-1} + b(s^{n-1} + \cdots + 1) + c + as^n - as^{n-1} + bs^{n+1} \\ &= f_n(s) + s^{n-1}(bs^2 + as - a) \\ &= f_n(s) + g(s)s^{n-1}. \end{aligned} \tag{2.12}$$

Note that d is the unique positive root of function g. Then, using (2.12), we obtain for all $n \geq 1$

$$f_n(d) = f_{n-1}(d) = \cdots = f_1(d).$$

Let i be any fixed but arbitrary natural number. Then, we get

$$f_i(d) = \lim_{n\to\infty} f_n(d) = f_\infty(d) \leq f_\infty(s_\infty) = 0, \tag{2.13}$$

since, function f_∞ is increasing, and $d \leq s_\infty$ by hypothesis (2.4). It follows from the definition of the zeros s_n and (2.13) that

$$d \leq s_n \text{ for all } n \geq 1. \tag{2.14}$$

Polynomials f_n are increasing which together with (2.14) imply

$$f_n(d) \leq f_n(s_n) = 0.$$

In particular

$$f_\infty(d) = \lim_{n\to\infty} f_n(d) \leq 0.$$

Hence, estimate (2.7) holds.

We then get from (2.12), and (2.14)

$$f_{n+1}(s_{n+1}) = f_n(s_{n+1}) + g(s_{n+1})s_{n+1}^{n-1}\eta$$

or

$$f_n(s_{n+1}) \leq 0, \tag{2.15}$$

since $f_{n+1}(s_{n+1}) = 0$, and $g(s_{n+1})s_{n+1}^{n-1}\eta \geq 0$, which imply

$$s_{n+1} \leq s_n \qquad (n \geq 1).$$

Sequence $\{s_n\}$ is non-increasing, bounded below by zero, and as such it converges to s^*.

We shall show $s_\infty \leq s^*$. Using (2.12), we have

$$\begin{aligned} f_{i+1}(s_i) &= f_i(s_i) + g(s_i)s_i^{i-1}\eta \\ &= g(s_i)s_i^{i-1}\eta \geq 0, \end{aligned}$$

so

$$f_{i+2}(s_i) = f_{i+1}(s_i) + g(s_i)s_i^{i}\eta \geq 0.$$

If, $f_{i+m}(s_i) \geq 0, m \geq 0$, then

$$f_{i+m+1}(s_i) = f_{i+m}(s_i) + g(s_i)s_i^{i+m-1}\eta \geq 0.$$

Hence, by the definition of function f_∞, we get

$$f_\infty(s_\infty) \geq 0 \text{ for all } n \geq 1.$$

But we also have $f_\infty(0) = b + c < 0$. That is $s_\infty \leq s_n$ for all $n \geq 1$, and consequently $s_\infty \leq s^*$.

□

Set $a = L\eta, b = 2L_0\eta$, and $c = -2$ in Theorem 2.1. It is simple algebra to show that conditions (2.4), and (2.5) reduce to (2.16) in the majorizing lemma that follows.

Lemma 2.2 Assume there exist constants $L_0 \geq 0, L \geq 0$, and $\eta \geq 0$, with $L_0 \geq L$, such that

$$q_0 = \overline{L}\eta \begin{cases} \leq \frac{1}{2} & \text{if} \quad L_0 \neq 0, \\ \\ < \frac{1}{2} & \text{if} \quad L_0 = 0. \end{cases} \tag{2.16}$$

where,

$$\overline{L} = \frac{1}{8}\left(L + 4L_0 + \sqrt{(L^2 + 8L_0L)}\right). \tag{2.17}$$

Then, sequence $\{t_k\}$ $(k \geq 0)$ given by

$$t_0 = 0, t_1 = \eta, t_{k+1} = t_k + \frac{L(t_k - t_{k+1})^2}{2(1 - L_0 t_k)} \quad (k \geq 1), \tag{2.18}$$

is nondecreasing, bounded above by t^{**}, and converges to its unique least upper bound $t^* \in [0, t^{**}]$, where

$$t^{**} = \frac{2\eta}{2 - \delta}, \tag{2.19}$$

$$\delta = \frac{4L}{L + \sqrt{L^2 + 8L_0L}} < 2 \text{ for } L_0 \neq 0. \tag{2.20}$$

Moreover the following estimates hold

$$L_0 t^* \leq 1, \tag{2.21}$$

$$0 \leq t_{k+1} - t_k \leq \frac{\delta}{2}(t_k - t_{k-1}) \leq \cdots \leq \left(\frac{\delta}{2}\right)^k \eta, \quad (k \geq 1), \tag{2.22}$$

$$t_{k+1} - t_k \leq \left(\frac{\delta}{2}\right)^k (2q_0)^{2^k - 1}\eta, \quad (k \geq 0), \tag{2.23}$$

$$0 \leq t^* - t_k \leq \left(\frac{\delta}{2}\right)^k \frac{(2q_0)^{2^k - 1}}{1 - (2q_0)^{2^k}}, \quad (2q_0 < 1), \quad (k \geq 0), \tag{2.24}$$

Proof. We shall show using induction on k that for all $k \geq 0$

$$L(t_{k+1}-t_k)+\delta L_0 t_{k+1} < \delta, \tag{2.25}$$

$$0 < t_{k_1}-t_k, \tag{2.26}$$

$$L_0 t_{k+1} < 1, \tag{2.27}$$

and

$$0 < t_{k+1} < t^{**}. \tag{2.28}$$

Estimates (2.25) - (2.28) hold true for $k=0$ by the initial condition $t_1 = \eta$, and hypothesis (2.16). It then follows from (2.18) that

$$0 < t_2 - t_1 \leq \frac{\delta}{2}(t_1 - t_0) \text{ and } t_2 \leq \eta + \frac{\delta}{2}\eta = \frac{2+\delta}{2}\eta < t^{**}.$$

Let us assume estimates (2.25) - (2.28) hold true for all integer value n $n \leq k$ $(n \geq 0)$. We also get

$$\begin{aligned} t_{k+1} &\leq t_k + \frac{\delta}{2}(t_k - t_{k-1}) \\ &\leq t_{k-1} + \frac{\delta}{2}(t_{k-1} - t_{k-2}) + \frac{\delta}{2}(t_k - t_{k-1}) \\ &\leq \eta + \left(\frac{\delta}{2}\right) + \cdots + \left(\frac{\delta}{2}\right)^k \eta \\ &= \frac{1-\left(\frac{\delta}{2}\right)^{k+1}}{1-\frac{\delta}{2}}\eta \\ &< \frac{2\eta}{2-\delta} = t^{**}. \end{aligned} \tag{2.29}$$

We have

$$L(t_{k+1}-t_k)+\delta L_0 t_{k+1} \leq L\left(\frac{\delta}{2}\right)^k \eta + L_0 \delta \frac{1-\left(\frac{\delta}{2}\right)^{k+1}}{1-\frac{\delta}{2}}\eta. \tag{2.30}$$

In view of estimate (2.30), (2.25) holds, if

$$\left\{ L\left(\frac{\delta}{2}\right)^n + \delta L_0 \frac{1-\left(\frac{\delta}{2}^{n+1}\right)}{1-\frac{\delta}{2}} \right\} \eta \leq \delta. \tag{2.31}$$

Estimate (2.31) motivates us to define for $s = \frac{\delta}{2}$, the sequence $\{f_n\}$ of polynomials on $[0, \infty)$ by

$$f_n(s) = \left(Ls^{n-1} + 2L_0(1 + s + s^2 + \cdots + s^n)\right)\eta - 2. \tag{2.32}$$

In view of Theorem 2.1, the induction for (2.25) - (2.28) is completed.

Hence, sequence $\{t_n\}$ is non-decreasing, bounded above by t^{**}, and as such that it converges to its unique least upper bound t^*. The induction is completed for (2.21), and (2.22).
If $L_0 = 0$, then (2.21) holds trivially. In this case, $L > 0$, an induction argument shows that

$$t_{k+1} - t_k = \frac{2}{L}(2q_0)^{2^k} \quad (k \geq 0),$$

and therefore

$$t_{k+1} = t_1 + (t_2 - t_1) + \cdots + (t_{k+1} - t_k) = \frac{2}{L}\sum_{m=0}^{k}(2q_0)^{2^m},$$

and

$$t^* = \lim_{k\to\infty} t_k = \frac{2}{L}\sum_{m=0}^{k}(2q_0)^{2^k}.$$

Clearly, this series converges, since $k \leq 2^k, 2q_0 < 1$, and is bounded above by the number

$$\frac{2}{L}\sum_{m=0}^{k}(2q_0)^{2^k} = \frac{4}{L(2 - L\eta)}.$$

If $L = 0$, then,
since, $0 \leq L_0 \leq L$, we deduce $L_0 = 0$, and $t^* = t_k = \eta(k \geq 1)$.
In the rest of the proof, we assume that $L_0 > 0$. In order for us to later complete the induction for (2.23), we first need to show the estimate.

$$\frac{1 - \left(\frac{\delta}{2}\right)^{k+1}}{1 - \frac{\delta}{2}} \leq \frac{1}{L_0}\left(1 - \left(\frac{\delta}{2}\right)^{k-1}\frac{L}{4\overline{L}}\right) \quad (k \geq 1). \tag{2.33}$$

For $k = 1$, (2.33) becomes

$$\left(1 + \frac{\delta}{2}\right)\eta \leq \frac{4\overline{L} - L}{4\overline{L}L_0}$$

or

$$\left(1+\frac{2L}{L+\sqrt{L^2+8L_0L}}\right)\eta \leq \frac{4L_0-L+\sqrt{L^2+8L_0L}}{L_0(4L_0+L+\sqrt{L^2+8L_0L})}$$

In view of (2.16), it suffices to show,

$$\frac{L_0(4L_0+L+\sqrt{L^2+8L_0L})(3L+\sqrt{L^2+8L_0L})}{(L+\sqrt{L^2+8L_0L})(4L_0-L+\sqrt{L^2+8L_0L})} \leq 2\overline{L},$$

which is true as equality.

Let us now assume estimate (2.33) is true for all integers small or equal to k. We must show (2.33) holds for k replaced by $k+1$,

$$\frac{1-\left(\frac{\delta}{2}\right)^{k+2}}{1-\frac{\delta}{2}}\eta \leq \frac{1}{L_0}\left(1-\left(\frac{\delta}{2}\right)^{k}\frac{L}{4\overline{L}}\right) \quad (k \leq 1).$$

or

$$\left(1+\frac{\delta}{2}+\left(\frac{\delta}{2}\right)^{2}+\cdots+\left(\frac{\delta}{2}\right)^{k+1}\right)\eta \leq \frac{1}{L_0}\left(1-\left(\frac{\delta}{2}\right)^{k}\frac{L}{4\overline{L}}\right) \quad (k \leq 1). \tag{2.34}$$

By the induction hypothesis to show (2.34), it suffices

$$\frac{1}{L_0}\left(1-\left(\frac{\delta}{2}\right)^{k-1}\frac{L}{4\overline{L}}\right)\eta \leq \frac{1}{L_0}\left(1-\left(\frac{\delta}{2}\right)^{k}\frac{L}{4\overline{L}}\right)$$

or

$$\left(\frac{\delta}{2}\right)^{k+1}\eta \leq \frac{1}{L_0}\left(\left(\frac{\delta}{2}\right)^{k-1}-\left(\frac{\delta}{2}\right)^{k}\right)\frac{L}{4\overline{L}}$$

or

$$\delta^2\eta \leq \frac{L(2-\delta)}{2\overline{L}L_0}$$

In view of (2.16) it suffices to show

$$\frac{2\overline{L}L_0\delta^2}{L(2-\delta) \leq 2\overline{L},}$$

which holds an equality by the choice of δ given by (2.20). That completes the induction for estimates (2.33).

We shall show (2.23) using induction on $k \geq 0$: estimate (2.23) is true for $k = 0$ by (2.16), (2.18), and (2.20). The inductive argument later in the proof requires the second base case of $k = 1$. In order for us to show estimate (2.23) for $k = 1$, since

$$t_2 - t_1 = \frac{L(t_1 - t_0)^2}{2(1 - L_0 t_1)}$$

it suffices

$$\frac{L\eta^2}{2(1 - L_0\eta)} \leq \delta \overline{L} \eta^2$$

or

$$\frac{L}{1 - L_0\eta} \leq \frac{8\overline{L}L}{L + \sqrt{L^2 + 8L_0L}} (\eta \neq 0)$$

or

$$\eta \leq \frac{1}{L_0}\left(1 - \frac{L + \sqrt{L^2 + 8L_0L}}{8\overline{L}}\right) \quad (L_0 \neq 0, L \neq 0).$$

But by (2.16)

$$\eta \leq \frac{4}{L + 4L_0 + \sqrt{L^2 + 8L_0L}}.$$

It then suffices to show

$$\frac{4}{L + 4L_0 + \sqrt{L^2 + 8L_0L}} \leq \frac{1}{L_0}\left(1 - \frac{L + \sqrt{L^2 + 8L_0L}}{8\overline{L}}\right)$$

or

$$\frac{L + \sqrt{L^2 + 8L_0L}}{8\overline{L}} \leq 1 - \frac{4L_0}{L + 4L_0 + \sqrt{L^2 + 8L_0L}}$$

or

$$\frac{L + \sqrt{L^2 + 8L_0L}}{8\overline{L}} \leq \frac{L + \sqrt{L^2 + 8L_0L}}{L + 4L_0 + \sqrt{L^2 + 8L_0L}},$$

which is true as equality by (2.17).
Let us assume (2.34) holds for all integers smaller or equal to k. We shall show (2.34) holds for k replaced by $k + 1$.

Using (2.18), and the induction hypothesis, we have in turn

$$\begin{aligned}
t_{k+2}-t_{k+1} &= \frac{L}{2(1-L_0t_{k+1})}(t_{k+1}-t_k)^2 \\
&\leq \frac{L}{2(1-L_0t_{k+1})}\left(\left(\frac{\delta}{2}\right)^k (2q_0)^{2^k-1}\eta\right)^2 \\
&\leq \frac{L}{2(1-L_0t_{k+1})}\left(\left(\frac{\delta}{2}\right)^{k-1} (2q_0)^{-1}\eta\right)\left(\left(\frac{\delta}{2}\right)^{k+1} (2q_0)^{2^{k+1}-1}\eta\right) \\
&\leq \left(\frac{\delta}{2}\right)^{k+1} (2q_0)^{2^{k+1}-1}\eta,
\end{aligned}$$

since,

$$\frac{L}{2(1-L_0t_{k+1})}\left(\left(\frac{\delta}{2}^{k-1}\right)(2q_0)^{-1}\eta\right) \leq 1, \quad (k \geq 1). \tag{2.35}$$

Indeed, we can show instead of (2.35)

$$t_{k+1} \leq \frac{1}{L_0}\left(1-\left(\frac{\delta}{2}\right)^{k-1}\frac{L}{4\overline{L}}\right),$$

which is true by (2.33), since, in view of (2.22), and the induction hypothesis

$$\begin{aligned}
t_{k+1} &\leq t_k + \frac{\delta}{2}(t_k - t_{k-1}) \\
&\leq t_1 + \frac{\delta}{2}(t_1 - t_0) + \cdots + \frac{\delta}{2}(t_k - t_{k-1}) \\
&\leq \eta + \left(\frac{\delta}{2}\right)\eta + \cdots + \left(\frac{\delta}{2}\right)^k \eta \\
&= \frac{1-\left(\frac{\delta}{2}\right)^{k+1}}{1-\frac{\delta}{2}}\eta \\
&\leq \frac{1}{L_0}\left(1-\left(\frac{\delta}{2}\right)^{k-1}\frac{L}{4\overline{L}}\right).
\end{aligned}$$

That completes the induction for estimate (2.23).

In view of (2.23), we obtain in turn for $2q_0 < 1$, and $j \geq k$.

$$\begin{aligned} t_{j+1}-t_k &= (t_{j+1}-t_j)+(t_j-t_{j-1})+\cdots+(t_{k+1}-t_k) \\ &\leq \left(\left(\frac{\delta}{2}\right)(2q_0)^{2^j-1}+\left(\frac{\delta^{j-1}}{2}\right)(2q_0)^{2^j-1}+\cdots+\left(\frac{\delta}{2}\right)(2q_0)^{2^k-1}\right)\eta \\ &\leq \left(1+(2q_0)^{2k}+\left((2q_0)^{2^k}\right)^2+\cdots\right)\left(\frac{\delta}{2}\right)^k(2q_0)^{2^k-1}\eta \\ &= \left(\frac{\delta}{2}\right)^k\frac{(2q_0)^{2^k-1}\eta}{1-(2q_0)^{2^k}}. \end{aligned} \tag{2.36}$$

Estimate (2.24) follows from (2.36) by letting $j \to \infty$.

□

3. Applications and Examples

As a first application, we show how to locate a root of a polynomial f_n $(n \geq 2)$, using, say e.g. s_{n-1}, and s_∞.

Application 3.1 Let $a = b = 1, c = -3$, and $n = 2$. We obtain using (2.1) - (2.3), and (2.10)

$$f_1(s) = s-1, f_2(s) = s^2+2s-2, s_1 = 1, s_\infty = \frac{2}{3}, d = 0.618033989.$$

Conditions (2.4), and (2.5) become

$$0.618033989 < \frac{2}{3},$$

and

$$-1 < 0, \text{ respectively.}$$

Hence, the conclusion of Theorem 2.1 hold. In particular, we know $k_2 \in (s_\infty, s_1)$. Actual direct computation justifies the theoretical claim, since

$$s_2 = \sqrt{3}-1 = 0.732050808 \in \left(\frac{2}{3}, 1\right).$$

Application 3.2 As a second application, we show how to use Theorem 2.1 to derive sufficient convergence conditions for scalar majorizing sequences of certain popular iterative methods such as Newton's method

$$x_{n+1} = x_n - F'(x_n)^{-1}F(x_n) \;\; (n \geq 0), \;\; (x_0 \in D), \tag{3.1}$$

where, F is a Fréchet-differentiable operator defined on a convex subset D of Banach space X with values in a Banach space Y.

Let us consider the famous Kantorovich hypotheses for solving nonlinear equations [3]

(K) $x_0 \in D$, with $F'(x_0)^{-1} \in L(Y,X)$,

$$\|F'(x_0)F(x_0)\| \leq \eta,$$
$$\|F'(x_0)^{-1}\Big(F'(x) - F'(y)\Big)\| \leq \|x-y\| \text{ for all } x,y \in D,$$
$$q_k = L\eta \leq \frac{1}{2}, \tag{3.2}$$
$$\overline{U}(x_0,r) = \{x \in X \mid \|x-x_0\| \leq r\} \subseteq D,$$

for

$$r = \frac{1-\sqrt{1-2q_k}}{l}.$$

Under the (K) hypotheses, the Newton-Kantorovich method converges quadratically (if $2q_k < 1$) to a unique solution $x^* \in \overline{U}(x_0,r)$ of equation $F(x) = 0$.

Moreover, scalar iteration $\{v_n\}$ $(n \geq 0)$, given by

$$v_0 = 0, v_1 = \eta, v_{n+2} = v_{n+1} + \frac{L(v_{n+1}-v_n)^2}{2(1-Lv_{n+1})},$$

is a majorizing sequence for $\{x_n\}$ in the sense that for all $n \geq 0$

$$\|x_{n+1} - x_n\| \leq v_{n+1} - v_n$$

and

$$\|x_n - x^*\| \leq r - v_n.$$

Let us consider our hypotheses

(A) $x_0 \in D$, with $F'(x_0)^{-1} \in L(Y,X)$,

$$\|F'(x_0)^{-1}F(x_0)\| \leq \eta,$$
$$\|F'(x_0)^{-1}\Big(F'(x) - F'(y)\Big)\| \leq L\|x-y\|,$$
$$\|F'(x_0)^{-1}\Big(F'(x) - F'(x_0)\Big)\| \leq L_0\|x-x_0\| \text{ for all } x,y \in D,$$
$$q_0 \leq \frac{1}{2},$$
$$\overline{U}(x_0,t^*) \subseteq D \ \ (\text{or } \overline{U}(x_0,t^{**}) \subseteq D).$$

Note that the center-lipschitz condition is not an additional hypothesis, since, in practice the computation of L requires that of L_0.

The scalar iteration $\{t_n\}$ is a finer majorizing sequence for $\{x_n\}$ that $\{v_n\}$ provided that $L_0 < L$ [1], [2], [3].

Remark 3.3 In general

$$L_0 \leq L$$

holds, and $\overline{L}L_0$ can be arbitrary large [1] - [7].

if $L = L_0$. Otherwise (2.16) is weaker that (3.2). Moreover the ratio $2q_0$ is also smallest than $2q_k$.

That is, (2.16) can always replace (3.2) in the Newton-Kantorovich theorem [3].

Hence, the applicability of Newton's method has been extended.

Example 3.4 Define the scalar function F by $F(x) = c_0x + c_1 + c_2 \sin e^{c_3x}$, $x_0 = 0$, where $c_i, i = 1,2,3$ are given parameters. Then it can easily be seen that for c_3 large and c_2 sufficiently small, $\frac{L}{L_0}$ can be arbitrarily large. That is (2.16) may be satisfied but not (3.2).

Example 3.5 Let $X = Y = \mathbb{R}^2$, be equipped with the max-norm, $x_0 = (1,1)^T$, $U_0 = \{x \mid \|x-x_0\| \leq 1-\beta\}, \beta \in \left[0, \frac{1}{2}\right)$, and define function F on U_0 by

$$F(x) = (w^3 - \beta, z^3 - \beta), \quad x = (w,z)^T. \tag{3.3}$$

The Fréchet-derivative of operator F is given by

$$F'(x) = \begin{bmatrix} 3v^2 & 0 \\ 0 & 3z^2 \end{bmatrix} \tag{3.4}$$

Using our hypotheses, we get

$$\eta = \frac{1}{3}(1-\beta), L_0 = 3-\beta, \text{ and } L = 2(2-\beta).$$

The Kantorovich condition (3.2) is violated, since

$$\frac{4}{3}(1-\beta)(2-\beta) > 1 \text{ for all } \beta \in \left[0, \frac{1}{2}\right)$$

Hence, there is no guarantee that Newton's method (3.1) converges to $x^* = (\sqrt[3]{\beta}, \sqrt[3]{\beta})^T$, starting at x_0.

However, our condition (2.16) is true for all $\beta \in I = \left[0.450339002, \frac{1}{2}\right)$.

Hence, our results apply to solve equation (2.2) for all $\beta \in I$. Other applications where $L_0 < L$ can be found in [1], [2].

Remark 3.6 Define scalar sequence $\{\alpha_n\}$ by

$$\alpha_0 = 0, \alpha_1 = \eta, \alpha_{n+2} = \alpha_n + \frac{L_1(\alpha_{n+1}-\alpha_n)^2}{2(1-L_0\alpha_{n+1})} \quad (n \geq 0), \tag{3.5}$$

where,

$$L_1 = \begin{cases} L_0, & \text{if} \quad n = 0 \\ L, & \text{if} \quad n > 0. \end{cases}$$

It follows from (2.18), and (3.5) that $\{\alpha_n\}$ converges under the hypotheses of Lemma (2.2.). Note that, under the (K) hypotheses, $\{\alpha_n\}$ is a finer majorizing sequence for $\{x_n\}$ than $\{t_n\}$. More precisely, we have the following estimates (if $L_0 < L$):

$$\|x_{n+1} - x_n\| \leq \alpha_{n+1} - \alpha_n < t_{n+1} - t_n < v_{n+1} - v_n, \quad (n \geq 1),$$

$$\|x_n - x^\star\| \leq \alpha^* - \alpha_n < t^* - t_n < v^* - v_n, \quad (n \geq 0),$$

$$\alpha_n < t_n < v_n, \quad (n \geq 2),$$

and

$$\alpha^* \leq t^* \leq r,$$

where,

$$\alpha^* = \lim_{n \to \infty} \alpha_n.$$

Remark 3.7 Our new technique of recurrent functions can be used on other Newton–type methods [1]-[7], so we can obtain the similar improvements as in the case of Newton's method. These advantages are generalized, since we use the (needed, and more precise than the Lipschitz) center-Lipschitz condition for the computation of the upper bounds of the norms $\|F'(x_n)^{-1}F'(x_0)\|$. This modification leads to more precise majorizing sequences, which in turn motivate the introduction of the recurrent functions.

References

[1] Amat S., Busquier S., Plaza S., Review of some iterative root-finding methods from a dynamical point of view, *Scientia Series A: Mathematical Sciences* 10 (2004) 3–35.

[2] Argyros, I.K., *Convergence and Applications of Newton–type iterations*, Springer–Verlag Pub., New York, 2008.

[3] Atkinson, K., *Elementary Numerical Analysis*, Wiley, NY, 1993.

[4] Kantorovich, L.V., Akilov, G.P., *Functional Analysis*, Pergamon Press, Oxford, 1982.

[5] Magreñán Á. A., Different anomalies in a Jarratt family of iterative root-finding methods, *Applied Mathematics and Computation* 233 (2014), 29–38.

[6] Magreñán Á. A., Cordero, A., Gutierrez, J. M. and Torregrosa, J.R., *Real qualitative behavior of a fourth-order family of iterative methods by using the convergence plane Mathematics and Computers in Simulation*, 105 (2014), 49-61.

[7] McNamee, J.M., *Numerical methods for roots of polynomials, part I*, 14, Elsevier, 2007.

Chapter 23

Hybrid High Convergence Order Iterative Methods

1. Introduction

We consider the iterative method defined for $n = 0, 1, 2, \ldots,$

$$x_{n+1} = x_n - F'(x_n)^{-1}F(x_n) + F'(x_n)^{-1}F(x_n)(I + A_n)^{-1}A_n, \tag{1.1}$$

where,

$$A_n = -\frac{1}{2}\frac{F''(x_n)F(x_n)}{F'(x_n)F'''(x_n)} + \frac{1}{6}\frac{F(x_n)^2}{F'''(x_n)F'(x_n)},$$

and x_0 is an initial point for solving the equation

$$F(x) = 0. \tag{1.2}$$

Here, $F : D \subset S \to S$ is a thrice continuously differentiable function and $S = \mathbb{R}$ or $S = \mathbb{C}$.

The aim of this chapter is to extend the applicability of a high convergence order iterative method (1.1).

2. Local Convergence

We present the local convergence of method (1.1) using some scalar functions and parameters. Let $\alpha > 0, \gamma > 0, L_0 > 0, L > 0$ and $M \geq 1$ be given parameters. Define functions p and h_p on the interval $[0, \frac{1}{L_0})$ by

$$p(t) = \frac{\alpha M t}{2(1 - L_0 t)} + \frac{\gamma M^2 t^2}{6(1 - L_0 t)}$$

and

$$h_p(t) = p(t) - 1.$$

We have that $h_p(0) = -1 < 0$ and $h_p(t) \to \infty$ as $t \to \frac{1}{L_0}^-$. It then follows from the intermediate value theorem that function h_p has zeros in the interval $(0, \frac{1}{L_0})$. Denote by r_p the smallest such zero. Moreover, define functions g and h on the interval $[0, r_p)$ by

$$g(t) = \frac{Lt}{2(1 - L_0 t)} + \frac{M p(t)}{(1 - L_0 t)(1 - p(t))}$$

and

$$h(t) = g(t) - 1.$$

Then, again we get $h(0) = -1 < 0$ and $h(t) \to \infty$ as $t \to r_p^-$. Denote by r the smallest zero of function h on the interval $(0, r_p)$. Then, we have that

$$0 < r < \frac{1}{L_0} \tag{2.1}$$

and for each $t \in [0, r)$

$$0 \leq p(t) < 1 \tag{2.2}$$

and

$$0 \leq g(t) < 1. \tag{2.3}$$

Denote by $U(y, \rho), \overline{U}(y, \rho)$ the open and closed balls in S with center $y \in S$ and of radius $\rho > 0$. Next, we present the local convergence analysis of method (1.1) using the preceding notation.

Theorem 2.1 Let $F : D \subseteq S \to S$ be a thrice continuously differentiable operator. Suppose that there exist $x^* \in D$ and $L_0 > 0$ such that for each $x \in D$

$$F(x^*) = 0, F'(x^*) \neq 0, \tag{2.4}$$

$$|F'(x^*)^{-1}\Big(F'(x) - F'(x^*)\Big)| \leq L_0|x_0 - x^*|. \tag{2.5}$$

Moreover, suppose that there exist $L > 0, M \geq 1, \alpha > 0$ and $\gamma > 0$ such that for each $x, y \in D_0 = D \cap U(x^*, \frac{1}{L_0})$

$$|F'(x^*)^{-1}\Big(F'(x) - F'(y)\Big)| \leq L|x_0 - y|, \tag{2.6}$$

$$|F'(x^*)^{-1}F'(x)| \leq M, \tag{2.7}$$

$$\sup_{x \in D_0} |\frac{F''(x)}{F'''(x)}| \leq \alpha \tag{2.8}$$

$$\sup_{x \in D_0} |\frac{F'(x^*)}{F'''(x)}| \leq \gamma \tag{2.9}$$

and

$$\overline{U}(x^*, r) \subseteq D, \tag{2.10}$$

where the radius of convergence r is defined previously. Then, sequence $\{x_n\}$ generated for $x_0 \in U(x^*, r) - \{x^*\}$ by method (1.1) is well defined in $U(x^*, r)$, remains in $U(x^*, r)$ for each $n = 0, 1, 2, \ldots$, and converges to x^*. Moreover, the following estimates hold

$$|x_{n+1} - x^*| \leq g(|x_n - x^*|)|x_n - x^*| \leq |x_n - x^*| < r, \tag{2.11}$$

where the g function is defined above Theorem (2.1). Furthermore , if $T \in [r, \frac{2}{L_0})$ the point x^* is the only solution of equation $F(x) = 0$ in $D_1 = D \cap \overline{U}(x^*, T)$.

Proof. We shall show that sequence $\{x_n\}$ is well defined, remains in $U(x^*, r)$ and converges to x^* so that estimate (2.11) is satisfied. Using hypothesis $x_0 \in U(x^*, r) - \{x^*\}$, the definition of r and (2.5), we have in turn that

$$|F'(x^*)^{-1}\Big(F'(x_0) - F'(x^*)\Big)| \leq L_0|x_0 - x^*| \leq L_0 r < 1. \tag{2.12}$$

It follows from (2.12) and the Banach lemma on invertible operators [1] that $F'(x_0) \neq 0$ and

$$|F'(x_0)^{-1}F'(x^*)| \leq \frac{1}{1 - L_0|x_0 - x^*|}. \tag{2.13}$$

We can write by (2.4) that

$$F(x_0) = F(x_0) - F(x^*) = \int_0^1 (F'(x^* + \theta(x_0 - x^*))(x_0 - x^*)d\theta. \tag{2.14}$$

Notice that $|x^* + \theta(x_0 - x^*) - x^*| = \theta|x_0 - x^*| < r$, so $x^* + \theta(x_0 - x^*) \in U(x^*, r)$ for each $\theta \in [0, 1]$. Then, by (2.7) and (2.14), we get that

$$|F'(x^*)^{-1}F(x_0)| \leq M|x_0 - x^*|. \tag{2.15}$$

Next, we must show that $A_0 \neq -1$. In view of (2.1), (2.2), (2.8), (2.9), (2.13) and (2.15) we obtain in turn that

$$\begin{aligned}
|A_0| &\leq \frac{1}{2}|\frac{F''(x_0)}{F'''(x_0)}||F'(x_0)^{-1}F'(x^*)||F'(x^*)^{-1}F(x_0)| \\
&+ \frac{1}{6}|\frac{F'(x^*)}{F'''(x)}||F'(x^*)^{-1}F(x_0)|^2|F'(x_0)^{-1}F'(x^*)| \\
&\leq \frac{\alpha M|x_0 - x^*|}{2(1 - L_0|x_0 - x^*|)} + \frac{\gamma M^2|x_0 - x^*|}{6(1 - L_0|x_0 - x^*|)} \\
&= p(|x_0 - x^*|) < p(r) < 1,
\end{aligned} \tag{2.16}$$

so

$$|(I + A_0)^{-1}| \leq \frac{1}{1 - p(|x_0 - x^*|))}, \tag{2.17}$$

and x_1 is well defined by method (1.1) for $n = 0$. Using (2.1), (2.3), (2.4), (2.6), (2.13), (2.15) - (2.17), we obtain in turn that

$$\begin{aligned}
|x_1 - x^*| &= |x_0 - x^* - F'(x_0)^{-1}F(x_0) + F'(x_0)^{-1}F(x_0)(I + A_0)^{-1}A_0| \\
&\leq |F'(x_0)^{-1}F'(x^*)||\int_0^1 F'(x^*)^{-1}\Big(F'(x^* + \theta(x_0 - x^*)) - F'(x_0)\Big)(x_0 - x^*)d\theta| \\
&+ |F'(x_0)^{-1}F'(x^*)||F'(x^*)^{-1}F(x_0)||(I + A_0)^{-1}||A_0| \\
&\leq \frac{L|x_0 - x^*|^2}{2(1 - L_0|x_0 - x^*|)} + \frac{Mp(|x_0 - x^*|)|x_0 - x^*|}{(1 - L_0|x_0 - x^*|)(1 - p(|x_0 - x^*|))} \\
&= g(|x_0 - x^*|)|x_0 - x^*| \leq |x_0 - x^*| < r,
\end{aligned} \tag{2.18}$$

which shows (2.11) for $n = 0$ and $x_1 \in U(x^*, r)$. By simply replacing x_0, x_1 by x_k, x_{k+1} in the preceding estimates we arrive at (2.11). Using the estimate

$$|x_{k+1} - x^*| \leq c|x_k - x^*| < r, \tag{2.19}$$

where,

$$c = g(|x_0 - x^*|) \in [0, 1).$$

We deduce that

$$\lim_{k \to \infty} x_k = x^* \text{ and } x_{k+1} \in U(x^*, r).$$

Finally, to show the uniqueness part, let

$$T = \int_0^1 F'(y^* + \theta(y^* - x^*))d\theta$$

for $y^* \in D_0$ with $F(y^*) = 0$. We have by (2.5) and the definition of D_1 that

$$|F'(x^*)^{-1}\left(T - F'(x^*)\right)| \leq L_0|y^* - x^*| < 1. \tag{2.20}$$

Then, from identity

$$0 = F(y^*) - F(x^*) = T(y^* - x^*), \tag{2.21}$$

we conclude that $y^* = x^*$. □

Remark 2.2 The convergence order of method (1.1) can be determined using computing the computational order of convergence (COC) [1] given by

$$\xi = \frac{\ln(\frac{|x_{n+2} - x^*|}{|x_{n+1} - x^*|})}{\ln(\frac{|x_{n+1} - x^*|}{|x_n - x^*|})} \tag{2.22}$$

or the approximate computational order of convergence (ACOC) [1] given by

$$\xi^* = \frac{\ln(\frac{|x_{n+2} - x_{n+1}|}{|x_{n+1} - x_n|})}{\ln(\frac{|x_{n+1} - x_n|}{|x_n - x_{n-1}|})}. \tag{2.23}$$

3. Numerical Examples

Example 3.1 Let $D = [-1, 1]$ and define function f on D by

$$f(x) = e^x - 1.$$

We get $\alpha = 1$, $L_0 = e - 1, \gamma = L = M = e^{\frac{1}{L_0}}$. The convergence radius is $r = 0.12711701064323011$.

References

[1] Argyros, I. K., *Convergence and Applications of Newton–type methods*, Springer Verlag, New York, 2008.

[2] Argyros, I. K., Verma, R. U., The generalized Newton's methods and local convergence analysis, *PanAmerican Mathematical Journal*, 26, 2, (2016), 74-79.

[3] Stewart, J., Calculus, Vol. 1, Thomson Books/Cole, 2006.

[4] Verma, R. U., *Higher order Newton's method, Advances in Nonlinear Variational Inequalities*, 19, 2, (2016), 82-93.

[5] Verma, R. U.,Hybrid-type new generation Newton's methods with higher order convergence, *PanAmerican Mathematical Journal*, 26, 4, (2016), 100–104.

[6] Zeidler, E., *Nonlinear functional analysis and its applications I*, Springer-Verlag, New York, 1986.

Chapter 24

Hybrid Newton-Like Methods with High Order of Convergence

1. Introduction

Quadratic convergence of hybrid-type Newton's method defined for $n = 0, 1, 2, \ldots$,

$$x_{n+1} = x_n - \frac{2f(x_n)}{2f'(x_n) - f(x_n)}, \tag{1.1}$$

where x_0 is an initial point and f is a real-valued continuously differentiable function was proved in [4]. Later in [5], cubic convergence of (1.1) was proved using assumptions on the derivatives of f of order up to three. In this chapter, using the assumptions only on the first derivative of fm we proved the convergence of method (1.1).

2. Local Convergence

The local convergence analysis of method (1.1) was based on hypotheses reaching the third derivatives limit the applicability of method (1.1) to functions f that are at least thrice differentiable on some interval containing the solution α of the equation $f(x) = 0$. As a motivational example, let us consider function f

defined on $D = [-\frac{5}{2}, \frac{3}{2}]$ by

$$f(x) = \begin{cases} x^3 \ln x^2 + x^5 - x^4, & x \neq 0 \\ 0, & x = 0. \end{cases} \tag{2.1}$$

Then, we have $\alpha = 1$, and

$$f'(x) = 3x^2 \ln x^2 + 5x^4 - 4x^3 + 2x^2$$
$$f''(x) = 6x \ln x^2 + 20x^3 - 12x^2 + 10x$$

and

$$f'''(x) = 6 \ln x^2 + 60x^2 - 24x + 22.$$

So obviously function $f'''(x)$ is not bounded on D. Therefore, the results in [5] cannot apply to solve equation $f(x) = 0$. In this Section the local convergence analysis of method (1.1) is based only on the first derivative. This way we expand the applicability of method (1.1). Moreover, we avoid Taylor expansions requiring the hypotheses on the third derivative of function f. Furthermore, in [5] the initial guess x_0 should be close to α for the convergence of method (1.1). But, how close the initial guess should be for the convergence of method (1.1). Finally, we need computable error bounds on the distances $|x_n - \alpha|$ and uniqueness of the solution α results. These concerns are addressed in what follows. Notice that the convergence order of method (1.1) can be obtained by computing the computational order of convergence (COC) [1] given by

$$\xi = \frac{\ln\left(\frac{|x_{n+2} - \alpha|}{|x_{n+1} - \alpha|}\right)}{\ln\left(\frac{|x_{n+1} - \alpha|}{|x_n - \alpha|}\right)} \tag{2.2}$$

or the approximate computational order of convergence (ACOC) [1] given by

$$\xi^* = \frac{\ln\left(\frac{|x_{n+2} - x_{n+1}|}{|x_{n+1} - x_n|}\right)}{\ln\left(\frac{|x_{n+1} - x_n|}{|x_n - x_{n-1}|}\right)} \tag{2.3}$$

that do not require high order derivatives. Notice also that ξ^* does not even require the knowledge of the exact root α. We shall use some functions and parameters in our convergence analysis of method (1.1). Let $w_0 : [0, \infty) \to [0, \infty)$,

$w : [0,\infty) \to [0,\infty), v : [0,\infty) \to [0,\infty)$ be continuous, non-decreasing functions with $w_0(0) = w(0) = 0$. Define parameters ρ_0, ρ_1 by

$$\rho_0 = \sup\{t \geq 0 : w_0(t) < 1\},$$

$$\rho_1 = \sup\{t \geq 0 : p(t) < 1\},$$

where $p(t) = \frac{1}{2}(w_0(t) + \int_0^1 v(\theta t) t d\theta$ and set

$$\rho = \min\{\rho_0, \rho_1\}. \tag{2.4}$$

Moreover, define functions g and h on the interval $[0,\rho)$ by

$$g(t) = \frac{\int_0^1 w((1-\theta)t)d\theta}{1 - w_0(t)} + \frac{(\int_0^1 v(\theta t)d\theta)^2 t}{2(1 - w_0(t))(1 - p(t))} \tag{2.5}$$

and

$$h(t) = g(t) - 1. \tag{2.6}$$

We have that $h(0) = -1 < 0$, and $h(t) \to \infty$ as $t \to \rho^-$. It then, follows from the intermediate value theorem that function h has roots in $(0,\rho)$. Denote by r the smallest such root. Then, we have that for each $t \in [0,r)$

$$0 \leq w_0(t) < 1 \tag{2.7}$$

$$0 \leq p(t) < 1 \tag{2.8}$$

and

$$0 \leq g(t) < 1. \tag{2.9}$$

Let $U(y,a), \bar{U}(y,a)$ denote, respectively the open and closed intervals in $\mathbb{R}$ with center $y \in \mathbb{R}$ and of radius $a > 0$.

Next, we present the local convergence analysis of method (1.1) using the preceding notation.

Theorem 2.1 Let $f : D \subseteq \mathbb{R} \to \mathbb{R}$ be a continuously differentiable function. Suppose:there exist $\alpha \in D$ such that

$$f(\alpha) = 0 \text{ and } f'(\alpha) \neq 0; \tag{2.10}$$

there exist functions $w_0 : [0,\infty) \to [0,\infty), w : [0,\rho) \to [0,\infty), v : [0,\rho) \to [0,\infty)$ continuous and nondecreasing with $w_0(0) = w(0) = 0$ such that

$$|f'(\alpha)^{-1}(f'(x) - f'(\alpha))| \leq w_0(|x-\alpha|), \text{ for each } x \in D, \tag{2.11}$$

$$|f'(\alpha)^{-1}(f'(x) - f'(y))| \leq w(|x-y|), \text{ for each } x \in D_0, \tag{2.12}$$

$$|f'(\alpha)^{-1}f'(x)| \leq v(|x-\alpha|) \text{ for each } x \in D_0 \tag{2.13}$$

$$\bar{U}(\alpha, r) \subseteq D, \tag{2.14}$$

where the radius of convergence r is defined previously and $D_0 = D \cap U(\alpha, \rho)$. Then, sequence $\{x_n\}$ generated for $x_0 \in U(\alpha, r) - \{\alpha\}$ by method (1.1) is well defined in $U(\alpha, r)$, remains in $U(\alpha, r)$ for each $n = 0, 1, 2, \ldots$ and converges to α. Moreover, the following estimates hold

$$|x_{n+1} - \alpha| \leq g(|x_n - \alpha|)|x_n - \alpha| \leq |x_n - \alpha| < r, \tag{2.15}$$

where the g function is given by (2.5). Further, if there exists $R \geq r$ such that

$$\int_0^1 w_0(\theta R) d\theta < 1, \tag{2.16}$$

then the limit point α is the only solution of equation $f(x) = 0$ in $D_1 = D \cap \bar{U}(\alpha, R)$.

Proof. We shall show using mathematical induction that sequence $\{x_n\}$ generated by method (1.1) is well defined, remains in $U(\alpha, r)$ and converges to α so that estimate (2.15) is satisfied. Using hypothesis $x_0 \in U(\alpha, r) - \{\alpha\}$, (2.7) and (2.11), we have in turn that

$$|f'(\alpha)^{-1}(f'(\alpha) - f'(x_0))| \leq w_0(|\alpha - x_0|) \leq w_0(r) < 1. \tag{2.17}$$

It follows from (2.17) and the Banach lemma on invertible functions [1] that $f'(x_0) \neq 0$ and

$$|f'(x_0)^{-1}f'(\alpha)| \leq \frac{1}{1 - w_0(|\alpha - x_0|)}. \tag{2.18}$$

We can write by (2.10) that

$$f(x_0) = f(x_0) - f(\alpha) = \int_0^1 (f'(\alpha + \theta(x_0 - \alpha))(x_0 - \alpha) d\theta. \tag{2.19}$$

Notice that $|\alpha+\theta(x_0-\alpha)-\alpha| = \theta|x_0-\alpha| < r$, so $\alpha+\theta(x_0-\alpha) \in U(\alpha,r)$ for each $\theta \in [0,1]$. In view, of (2.13) and (2.19), we get that

$$|f'(\alpha)^{-1}f'(x_0)| \leq \int_0^1 v(\theta|x_0-\alpha|)d\theta|x_0-\alpha|. \tag{2.20}$$

Next, we need to show that $2f'(x_0)-f(x_0) \neq 0$. By (2.8), (2.10), (2.11) and (2.20) we get in turn that

$$\begin{aligned}
&|(2f'(\alpha))^{-1}[2f'(x_0)-f(x_0)+f(\alpha)-2f'(\alpha)]| \\
&\leq \frac{1}{2}[|f'(\alpha)^{-1}(f'(x_0)-f'(\alpha))|+|f'(\alpha)^{-1}(f(x_0)-f(\alpha))|] \\
&\leq \frac{1}{2}(w_0(|x_0-\alpha|)+\int_0^1 v(\theta|x_0-\alpha|)d\theta|x_0-\alpha|) \\
&= p(|x_0-\alpha|) \leq p(r) < 1,
\end{aligned} \tag{2.21}$$

so $2f'(x_0)-f(x_0) \neq 0$, and

$$|(2f'(x_0)-f(x_0) \neq 0.)^{-1}f'(\alpha)| \leq \frac{1}{2(1-p(|x_0-\alpha|))}. \tag{2.22}$$

We also have by (2.22) and method (1.1) for $n=0$ that x_1 is well defined. Using method (1.1) and (2.10), we can write the identity

$$\begin{aligned}
x_1-\alpha &= x_0-\alpha-f'(x_0)^{-1}f(x_0)+f'(x_0)^{-1}f(x_0)-2(2f'(x_0)-f(x_0))^{-1}f(x_0) \\
&= (x_0-\alpha-f'(x_0)^{-1}f(x_0)-f'(x_0)^{-1}f(x_0)(2f'(x_0)-f(x_0))^{-1}f(x_0).
\end{aligned} \tag{2.23}$$

Then, using (2.9), (2.10), (2.12), (2.18), (2.22) and (2.23), we obtain in turn that

$$\begin{aligned}
|x_1-\alpha| &\leq |f'(x_0)^{-1}f'(\alpha)||\int_0^1 f'(\alpha)^{-1}(f'(\alpha+\theta(x_0-\alpha))-f'(x_0))(x_0-\alpha)d\theta| \\
&\quad +|f'(x_0)^{-1}f'(\alpha)||(2f'(x_0)-f(x_0))^{-1}f(\alpha)||f'(\alpha)^{-1}f(x_0)|^2 \\
&\leq \frac{\int_0^1 w((1-\theta)|x_0-\alpha|)d\theta|x_0-\alpha|}{1-w_0(|x_0-\alpha|)} \\
&\quad + \frac{(\int_0^1 v(\theta|x_0-\alpha|)d\theta|x_0-\alpha|)^2}{2(1-w_0(|x_0-\alpha|))(1-p(|x_0-\alpha|))} \\
&= g(|x_0-\alpha|)|x_0-\alpha| \leq |x_0-\alpha| < r,
\end{aligned} \tag{2.24}$$

which shows (2.15) for $n = 0$ and $x_1 \in U(\alpha, r)$. By simply replacing x_0, x_1 by x_k, x_{k+1} in the preceding estimates we arrive at (2.15). Then, by the estimate

$$|x_{k+1} - \alpha| \leq \gamma(|x_k - \alpha| < r, \tag{2.25}$$

for $\gamma = g(|x_0 - \alpha|) \in [0, 1)$, we deduce that

$$\lim_{k \longrightarrow \infty} x_k = \alpha \text{ and } x_{k+1} \in U(\alpha, r).$$

Finally, to show the uniqueness part, let

$$T = \int_0^1 f'(\alpha + \theta(\alpha^* - \alpha))d\theta \text{ for } \alpha^* \in D_0$$

with $f(\alpha^*) = 0$. We have by (2.11) and (2.16) that

$$|f'(\alpha)^{-1}(T - f'(\alpha))| \leq \int_0^1 w_0(|\alpha - \alpha^*|)d\theta \leq \int_0^1 w_0(\theta R)d\theta < 1, \tag{2.26}$$

so $T \neq 0$. Then, from identity

$$0 = f(\alpha^*) - f(\alpha) = T(\alpha^* - \alpha), \tag{2.27}$$

we conclude that $\alpha^* = \alpha$. □

3. Numerical Examples

Example 3.1 Let $D = [-1, 1]$ and define function f on D by

$$f(x) = e^x - 1.$$

We get $w_0(t) = (e-1)t, w(t) = e^{\frac{1}{e-1}}t, v(t) = 2$. the convergence radius is $r = 3.058558860636854781$, so we choose $r = 1$ since $x \in D$.

Example 3.2 Returning back to the motivational example (2.1) and using (2.10)-(2.13), we can choose $w_0(t) = w(t) = 96.6629073t$ and $v(t) = \sup|f'(\alpha)^{-1}f'(x)| = 0.7272$ for $x \in D$ to obtain that the radius $r = 0.0054$. As already mentioned in Section 2, the results in [5] cannot apply, since $f'''(x)$ is unbounded on D. Hence, there is no guarantee under the results in [4] that method (1.1) converges to $\alpha = 1$. However, Theorem 2.1 guarantee the convergence of method (1.1) to α.

References

[1] Argyros, I. K., *Convergence and Applications of Newton–type methods*, Springer Verlag, New York, 2008.

[2] Argyros, I. K., Verma, R. U., The generalized Newton's methods and local convergence analysis, *PanAmerican Mathematical Journal*, 26, 2, (2016), 74-79.

[3] Stewart, J., *Calculus*, Vol. 1, Thomson Books/Cole, 2006.

[4] Verma, R. U., Higher order Newton's method, *Advances in Nonlinear Variational Inequalities*, 19, 2, (2016), 82-93.

[5] Verma, R. U., Hybrid-type new generation Newton's methods with higher order convergence, *PanAmerican Mathematical Journal*, 26, 4, (2016), 100–104.

[6] Zeidler, E., *Nonlinear functional analysis and its applications I*, Springer-Verlag, New York, 1986.

Glossary of Symbols

$\mathbb{R}^n$	Real n-dimensional space
$\mathbb{C}^n$	Complex n-dimensional space
$X \times Y, X \times X = X^2$	Cartesian product space of X and Y
$e^1, \ldots, e^n$	The coordinate vector of $\mathbb{R}^n$
$x = (x_1, \ldots, x_n)^T$	Column vector with component x_i
x^T	The transpose of x
$\{x_n\}^{n \geq 0}$	Sequence of point from X
$\|.\|$	Norm on X
$\|.\|_p$	L_p norm
$\lvert . \rvert$	Absolute value symbol
$/./$	Norm symbol of a generalized Banach space X
$U(x_0, R)$	Open ball $\{z \in X \mid \|x_0 - z\| < R\}$
$\overline{U}(x_0, R)$	Closed ball $\{z \in X \mid \|x_0 - z\| \leq R\}$
$U(R) = U(x_0, R)$	Ball centered at the zero element on X and of radius R
$U, \overline{U}$	Open, closed balls, respectively no particular reference to X, x_0 or R
I	Identity matrix operator
L	Linear operator
L^{-1}	Inverse
$M = \{m_{ij}\}$	Matrix $1 \leq i, J \leq n$
M^{-1}	Inverse of M
$\det M$ or $\lvert M \rvert$	Determinant of M
Σ	Summation symbol
Π	Product of factors symbol
$\int$	Integration symbol
$\in$	Element inclusion
$\subset, \subseteq$	Strict and non-strict set inclusion
$\forall$	For all

$\Rightarrow$	Implies
$\cup, \cap$	Union, intersection
$A - B$	Difference between set A and B
$F : D \subseteq X \to Y$	An operator with domain D included in X, and values in Y
$F'(x), F''(x)$	First, second Fréchet-derivatives of F evaluated at x

About the Authors

Ioannis K. Argyros
Professor (Advisor)
Department of Mathematical Sciences
Cameron University, Lawton, OK, US
Email: iargyros@cameron.edu

Dr. Ioannis K. Argyros is a professor in the Department of Mathematical Sciences, Cameron University. His expertise is in research for the theories of inequalities, operators, computational mathematics and iterative methods, especially in Banach Spaces. He has published more than 1000 papers (included by Science Citation). He has also published 32 mathematical books.

Samundra Regmi
Research Associate
Department of Mathematical Sciences
Cameron University, Lawton, OK, US
Email: samundra.regmi@cameron.edu

Samundra Regmi is a mathematics, computer science, and information technology student. In addition, he is serving as a Research Associate in the Department of Mathematical Sciences, Cameron University. His area of focus and specialization is in research for numerical analysis, scientific computing, applied and computational mathematics, cryptography and optimization.

Index

A

algorithm, 33, 52
applied mathematics, 1, 65, 283
approximation, 51, 113, 115, 154, 156

B

Banach lemma, 20, 36, 68, 87, 90, 100, 109, 113, 125, 128, 136, 140, 147, 153, 165, 175, 187, 197, 209, 219, 229, 239, 249, 259, 288, 296
Banach spaces, 29, 31, 51, 65, 303
bounds, 2, 5, 8, 16, 41, 75, 88, 102, 110, 117, 126, 138, 151, 158, 164, 174, 175, 209, 259, 281, 294

C

closed ball, 18, 34, 53, 65, 173, 186, 196, 207, 217, 227, 238, 258, 286, 301
computational cost, 9, 11, 29, 111, 138
convergence radius, 40, 54, 66, 290, 298

D

derivative, 16, 25, 26, 31, 32, 33, 35, 37, 39, 41, 43, 45, 47, 79, 81, 83, 96, 97, 104, 120, 121, 123, 125, 127, 129, 132, 133, 144, 160, 161, 162, 169, 171, 181, 183, 184, 191, 194, 203, 205, 206, 213, 215, 216, 223, 225, 226, 233, 235, 236, 243, 246, 253, 255, 256, 265, 280, 293, 294
differential equations, 39
divided difference, 32, 42, 161, 163, 183, 205, 207, 225, 228, 248

F

fixed point, 51, 63
functional analysis, 291, 299

H

Halley method, 66, 75, 81, 82, 169, 170, 181, 182, 191, 192, 203, 204, 213, 214, 223, 224, 233, 234, 243, 244, 253, 254, 265, 266

I

IMA, 95, 96, 103, 104, 119, 120, 131, 132, 143, 144, 159, 160
induction, 6, 8, 9, 19, 35, 54, 55, 58, 59, 71, 73, 91, 112, 114, 129, 141, 153, 155, 175, 176, 190, 200, 210, 219, 221, 229, 231, 239, 242, 249, 251, 259, 263, 272, 273, 274, 275, 276, 277, 296
iteration, 11, 52, 54, 71, 75, 82, 174, 187, 197, 208, 209, 215, 229, 239, 249, 259, 278, 279

K

Kantorovich theorem, 4, 8, 96, 104, 120, 132, 144, 160, 279

L

local convergence, 2, 16, 18, 29, 30, 32, 33, 34, 42, 52, 53, 56, 66, 70, 71, 76, 81, 82, 83, 85, 86, 89, 92, 93, 95, 96, 97, 99, 100, 102, 103, 104, 105, 107, 108, 111, 112, 119, 120, 121, 123, 124, 127, 131, 132, 133, 134, 138, 139, 143, 144, 145, 151, 152, 159, 160, 161, 162, 163, 164, 169, 171, 172, 173, 174, 181, 183, 184, 186, 187, 191, 193, 194, 197, 203, 205, 206, 207, 208, 213, 215, 216, 217, 218, 223, 225, 226, 227, 229, 231, 233, 235, 236, 237, 238, 239, 241, 243, 245, 246, 247, 248, 249, 251, 253, 255, 256, 258, 265, 286, 293, 294, 295

M

mathematical induction, 6, 19, 35, 54, 59, 112, 153, 175, 219, 229, 239, 249, 259, 296
mathematical modeling, 1, 15, 65

N

numerical analysis, 96, 104, 120, 132, 144, 160, 283, 303

O

optimization, 1, 303

P

parameters, 16, 33, 41, 52, 66, 76, 162, 172, 184, 279, 286, 294, 295

Q

quadratic, 267, 293

R

radius, 18, 19, 24, 34, 40, 47, 53, 54, 65, 66, 68, 70, 75, 77, 89, 93, 117, 127, 158, 163, 164, 173, 174, 185, 186, 195, 196, 207, 208, 217, 227, 238, 247, 248, 258, 286, 287, 290, 295, 296, 298, 301
radius of convergence, 18, 19, 24, 34, 66, 68, 75, 89, 93, 127, 163, 164, 174, 185, 195, 207, 208, 238, 247, 287, 296
root(s), 25, 98, 107, 267, 268, 269, 277, 283, 294, 295

S

scalar function, 33, 52, 76, 133, 134, 135, 279, 286
solution, 1, 2, 5, 8, 15, 16, 19, 25, 31, 33, 34, 35, 43, 47, 51, 52, 54, 57, 58, 60, 65, 66, 67, 68, 71, 76, 78, 84, 85, 86, 88, 90, 93, 94, 98, 99, 100, 102, 106, 107, 108, 110, 111, 112, 122, 123, 124, 126, 127, 128, 134, 135, 136, 138, 140, 146, 147, 151, 152,162, 164, 172, 173, 174, 184,

185, 187, 194, 195, 200, 206, 207, 208, 216, 217, 219, 226, 227, 229, 236, 237, 238, 239, 246, 247, 249, 256, 257, 258, 259, 278, 287, 293, 294, 296

T

Taylor expansion, 32, 33, 294
Taylor series, 112, 153

V

variables, 96, 104, 120, 132, 144, 160
vector, 301

Related Nova Publications

The Life and Times of the World's Most Famous Mathematicians

Authors: Ann F. Varela and Michael F. Shaughnessy

Series: Mathematics Research Developments

Book Description: Arithmetic, math, algebra, geometry, trigonometry, calculus and statistics are all part of the modern world, yet so few of us know where these fields of study originated. In this book, the lives and discoveries of various mathematicians will be explored and examined, so as to provide some insight into various mathematical functions and operations.

Hardcover ISBN: 978-1-53613-975-4
Retail Price: $195

Single Variable Integral and Differential Calculus in a Nutshell with Elements of Critical Thinking

Authors: Ranis Ibragimov and Pirooz Mohazzabi

Series: Mathematics Research Developments

Book Description: This book presents a variety of calculus problems concerning different levels of difficulty with technically correct solutions and methodological steps that look also correct, but that have obviously wrong results (like 0 = 1).

Hardcover ISBN: 978-1-53614-047-7
Retail Price: $195

To see a complete list of Nova publications, please visit our website at www.novapublishers.com